T0327448

LACTATION AND THE MAMMARY GLAND

LACTATION AND THE MAMMARY GLAND

R. MICHAEL AKERS

Blackwell Publishing

R. Michael Akers, Ph.D., is Horace E. and Elizabeth F. Alphin Professor, Department of Dairy Science, Virginia Polytechnic Institute and State University. He has received numerous awards for research excellence and has been extremely active as a speaker and scholarly author in lactation physiology.

© 2002 Iowa State Press
A Blackwell Publishing Company
All rights reserved

Blackwell Publishing Professional
2121 State Avenue, Ames, Iowa 50014

Orders: 1-800-862-6657
Office: 1-515-292-0140
Fax: 1-515-292-3348
Web site (secure): www.blackwellprofessional.com

First edition, 2002

Library of Congress Cataloging-in-Publication Data

Akers, R. Michael.
   Lactation and the Mammary Gland / R. Michael Akers.
      p. cm.
Includes bibliographic references and index.
   ISBN 0-8138-2992-5
1. Lactation. 2. Milk yield. 3. Dairy cattle.
I. Title.
QP246 .A39 2002
573.6'79—dc21

                                                        2001005961

Contents

Preface

Lactation can be considered from a number of perspectives. For the dairy industry adequate mammary development and differentiation of the mammary secretory cells of cows, goats, sheep, and camels to synthesize and secrete milk are imperative. For other mammalian animal enterprises not directly involved in harvesting of milk, lactation is nonetheless critical for the successful rearing of replacement animals to supply meat and fiber as well as for rearing many companion animals. For human welfare, nursing and care of infants and breast cancer are intimately tied to the mammary gland and lactation. The study of mammary gland development and function also provides research material for a variety of scientific disciplines, including cell biology, biochemistry, immunology, food chemistry, and endocrinology to name a few.

The goal of this book is to provide the basics for understanding mammary development and lactation as well as some appreciation of critical regulatory events. Toward this aim the book provides a core of information so that it may be used as a textbook to support undergraduate or graduate courses in lactation. The book also serves as a source for new researchers interested in gaining an overview of mammary development and lactation as well as a tool for established scientists in need of review. Lastly, the book is a valuable resource for professionals in the animal and dairy industry and those generally interested in this important topic.

I would like to thank my students and superb scientific colleagues for their hard work and skill in completing the research described in the book. I thank my wife Cathy Akers for her patience during the process of preparing the text as well as her love and support for the past 31 years!

LACTATION AND THE MAMMARY GLAND

Chapter 1
Overview of Mammary Development

As all biology students learn early in their schooling, a distinguishing characteristic of mammals is the presence of mammary glands. With some notable exceptions (e.g., bottle-fed human infants or early-weaned dairy calves) lactation is critical for survival of the neonate and, ultimately, the reproductive success of mammals. Regardless of the specific arrangement or number of mammary glands for a given mammal, milk synthesis and secretion require development of a functionally mature mammary gland. In reproductively competent animals the mature mammary gland consists of a teat or nipple, associated ducts that provide for passage of milk to the outside, and alveoli composed of epithelial secretory cells and supporting tissues. The epithelial cells are arranged to form the internal lining of the spherical alveoli, and the cells synthesize and secrete all milk. Secretions are stored within the internal space of the hollow alveoli and larger ducts between suckling episodes. Consequently, understanding lactation requires an appreciation of numerous life science topics, including histology and cytology, biochemistry, endocrinology, cardiovascular physiology, metabolism, and developmental biology to name a few. Clearly, the fully functioning mammary gland places dramatic demands on the physiology of the lactating mother.

Dramatic development of the mammary gland during gestation and subsequent differentiation of alveolar cells to allow onset of milk synthesis and secretion in precise correspondence with parturition is indeed a biological marvel. Mammary secretion, first as colostrum and subsequently as mature milk, provides the neonate with a spectrum of all the nutrients necessary for good health and early development. Nutritionally, milk of all mammals contains variable amounts of proteins, carbohydrates, and fats suspended in an aqueous medium, thus providing each of the major classes of nutrients. Although there are species differences in milk composition, having the birth of the neonate and functionality of the mammary gland coincide is obviously critical. Scientists' study of lactation and mammary development provides a rich resource for cell biologists, endocrinologists, nutritionists, cancer specialists, dairy specialists, and others. The goal of this text is to provide a core of information to assist in the learning or review of lactation and mammary development.

Evolutionary Aspects

Despite the lack of soft tissue fossils, comparisons of mammary development among various classes of mammals suggest that the ectoderm-derived mammary gland arose from sweat or sebaceous glands. Secretions associated with incubation and care of eggs may have been the precursors of milk-like secretions from brood patches on the abdominal or inguinal surface of early mammals. Reproductive tract secretions, like mammary colostrum, may also contain antibacterial substances. Indeed, mammary secretions contain lactoferrin, immunoglobulins, and other bactericidal components. Thus, the evolution of a skin-gland product to provide support for a newly hatched offspring in addition to protection of the egg prior to hatching is intuitively appealing. Support for this course of development comes from descriptions of mammary development in the primitive egg-laying mammals the monotremes. For example, the echidna has two mammary glands on either side of the abdomen. Each gland is arranged with clusters of lobules, each with a separate duct that opens on the surface of the skin in a small depression; however, there are no teats or nipples.

Hypotheses for the adaptive value of proto-lacteal secretions are based on thermoregulatory, antibiotic, behavioral, or nutritive functions. A reasonable conjecture is that lactation arose in endothermic, oviparous ancestors that exhibited at least some degree of maternal care. Early anatomists tried to define the origin of the mammary gland by classifying the secretion mechanism for the secretory cells. For example, sebaceous glands exhibit a holocrine mode of secretion in which cells are sloughed to become a part of the secretion. Sweat glands follow an apocrine mode of secretion in which only portions of the cells are lost so that individual cells are capable of periodic secretion. Other glands follow a meocrine mode of secretion in which products are secreted but the secretory cells remain intact. Mammary cells follow both apocrine and meocrine modes of secretion. To illustrate, as lipid droplets form in the cytoplasm of the cells, they progressively enlarge, migrate to the apical end of the cell, and protrude into the alveolar lumen until the membrane-bound droplets pinch off to become the butterfat of milk. Since the membrane surrounding the lipid droplet is derived from the plasma membrane of the cell, it is clear that a part of the cell is lost to become a part of the cellular secretion. This is an example of an apocrine mode of secretion. For secretion of specific milk proteins and milk sugar (lactose), these products are packaged into secretory vesicles in the Golgi apparatus. These vesicles both singly or in chains fuse with the apical plasma membrane and release their contents via exocytosis. Since only the secretory vesicle contents are lost from the cell, this mode of secretion is meocrine. The details for secretion patterns of mammary cells were not settled until mammary tissue from lactating mammals was studied with transmission electron microscopy in the early 1960s. It seems likely that the primitive mammary gland arose from a hybrid combination of both types of glandular cells (Blackburn et al., 1989).

For placental mammals, the number of mammary glands varies markedly among classes and species. However, among those studied to date, each mammary gland has a teat or nipple. It is nonetheless worth remembering that only a few of the known mammals have been studied. For example, lactation is not common in males, but development of small amounts of mammary tissue and limited secretion occurs. Lactation in males has been reported to occur spontaneously in humans and is most likely associated with pituitary dysfunction. However, anecdotal tales of "witch's milk" in male and female infants are not rare. Apparently, normal lactation also was recently reported in male wild fruit bats. Likely, the first reported lactation in a male ruminant was when Aristotle noted in his *Historia Animalium* that "a he-goat was milked by his dugs (teats) to such effect that cheese was made of the product." These examples, only serve to illustrate how little is known about mammary development and lactation in many mammals.

Species Variation

Given the variety of mammals and the environmental niches occupied, it is no surprise that there is much variation in number of mammary glands, location of mammary glands, and composition of secretions. Unlike common dairy species (cows, goats, or sheep) aquatic mammals, especially those in cold environments, produce milk very high in lipid content with relatively less lactose. High lipid content is essential for the suckling young to rapidly produce a layer of insulating fat to protect from the cold and to provide a source of metabolically derived water. This illustrates the relevance of lactation to provide a strategy for survival of offspring and reproductive success. Table 1.1 illustrates some of the variety in numbers and location of mammary glands in some common species.

The first indication for the presence of mammary development is a slight thickening of the ventrolateral ectoderm in the embryo at the time limb buds begin to lengthen. This thickening is variously referred to as the mammary band, streak, or line. Among species that have mammary glands along the entire ventral surface (i.e., rodents), a mammary line is evident from forelimb to hindlimb. For species having only pectoral glands, the mammary lines are confined to the thoracic area. The cells of the mammary line progressively condense through a series of somewhat arbitrary stages, the mammary crest and hillock, to develop mammary buds. A single clump or ball of cells is the precursor for an individual mammary gland. In the mouse, for example, each of five mammary buds on either side of the ventral midline is oriented at the location of the each of the future nipples.

Fundamentals of Structure and Function

It is now routine for herds of Holstein cows to average 305-day lactation yields of 13,000 kg of milk and for individual cows to produce much more milk. Such prodigious production of milk requires massive mammary glands and careful attention to the feeding and management of these impressive animals. While these facts illustrate the success of modern dairy husbandry, it is clear that milk production of the great whales is even more. Daily milk yield of many rodents can

Table 1.1 Variation in location, number, and nipple openings of mammary glands of some common species

Order	Common name	Position of glands: thoracic	abdominal	inguinal	Total glands	Opening per teat
Marsupialia	Red kangaroo		4		4	15
Marsupialia	Opossum		13		13	8
Carnivora	House cat	2	6		8	3–7
Carnivora	Domestic dog	2	6	2	10	8–14
Rodentia	House mouse	4	2	4	10	1
Rodentia	Norway rat	4	4	4	12	1
Lagomorpha	Rabbit	4	4	2	10	8–10
Cetacea	Whale			2	2	1
Proboscidea	Elephant	2			2	10–11
Perissodactyla	Horse			2	2	2
Artiodactyla	Cattle			4	4	1
Artiodactyla	Sheep			2	2	1
Artiodactyla	Goat			2	2	1
Artiodactyla	Pig	4	6	2	12	2
Primate	Human	2			2	15–25

Source. Adapted from Larson, 1985.

equal 10 percent of body weight. Whatever the rate of milk production for a particular mammal, all milk is produced by the secretory cells of the mammary alveolus. Consequently, to appreciate milk secretion, it is essential to understand the structure and function of the mammary gland.

Mammary glands, like all multicellular glands, are epithelial. Although the evolutionary origins are unclear, the mammary epithelium arises from the germinal ectoderm and the primitive mammary buds. In the cow, for example, the mammary bud appears at about day 40 of gestation. By day 80 the teat and primary sprout have formed. The primary sprout gives rise to the teat cistern. Secondary sprouts occur by day 90, and by day 100 the primary and secondary sprouts are canalized by apoptosis. At the time of birth, the teat, teat cistern, and gland cistern are formed. In most species, mammary structure at birth is similarly rudimentary. Figure 1.1 illustrates the epithelial portion of the mammary gland of a mouse at 1 week. The structure of the nipple and branches from a number of secondary sprouts are evident. For mammals without the mammary glands arranged into an udder, a teat or gland cistern is absent, but there is a nipple and a cluster of primary and secondary sprouts for each gland. Further growth of secondary

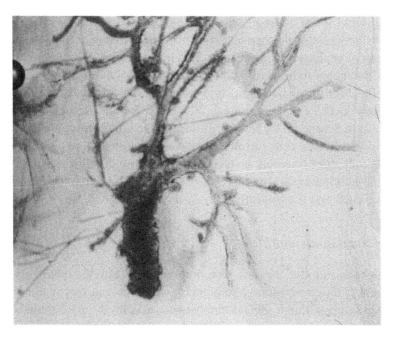

Fig. 1.1 Epithelial portion of the mammary gland of a 1-week-old female mouse. The fourth inguinal mammary gland was removed, mounted on a glass slide, fixed, defatted, and stained. The region surrounding the nipple (lower portion of the image) was photographed. The presence of a major galactophore associated with the nipple and branches is evident.

sprouts yields the major ducts that drain groups of alveoli (lobules) in the mature gland. In those species studied, development of the alveoli is usually restricted to pregnancy.

Epithelial glands follow several distinct patterns of development based on the arrangement of cells within the secreting unit of the glands. Simple glands have a duct that opens onto a surface. Usually cells that create the duct opening or neck are nonsecretory and serve as a passageway for products made deeper within the gland. The shape of the gland mimics the shape of tubes or rounded flasks called *alveoli* or *acini*. The presence of a single glandular unit denotes a simple gland. Depending on the shape of the secretory structure, the gland is classified as simple tubular or simple alveolar. By contrast, compound glands are branched with multiple secretory units opening into a duct. Again, depending on the specifics of the secretory units, glands are classified as compound tubular, alveolar, or tubulo-alveolar. Mammary glands are compound alveolar glands. However, clusters of alveoli and the ducts that drain them are arranged into units called *lobules*. The ducts that drain individual alveoli lead to progressively larger ducts, which connect with the nipple or teat openings. Within the alveolus, lining epithelial cells are a single cell layer thick (simple epithelium), but the nonsecretory ducts are stratified with two or more layers of cells. Creation of the lobulo-alveolar structure during gestation does not automatically lead to the onset of milk secretion by the alveolar cells. These cells must undergo progressive biochemical and structural differentiation to prepare the cells for onset of copious milk secretion at parturition. Figure 1.2 gives a histological view of mammary tissue during late pregnancy (2A) and lactation (2B) and a diagram of the three-dimensional structure for a fully developed alveolus and associated terminal duct of a lactating animal (2C). Major features include the secretory epithelial cells that make up the internal surface of the hollow alveolus, the star- or basket-like myoepithelial cells surrounding the alveolus, and the supporting vascular bed. The milk ejection reflex involves the oxytocin-induced contraction of the myoepithelial cells to increase internal pressure of the alveolus to force stored secretions into the ducts and subsequently to the teat or nipple.

Mammary Secretions and Milk

While it might be intuitive that any secretion obtained from the mammary gland is milk, in reality the composition of mammary secretions can vary markedly with physiological status within a species and between species during established lactation. As illustrated by the staining of lumen contents in Figure 1.2A, secretions begin to accumulate soon after the formation of alveoli during pregnancy. For example, in cows during first gestation small but variable volumes of secretions can be expressed from the mammary gland several months prior to parturition. Prepartum milking of dairy heifers is sometimes initiated during the month before parturition as a management technique to relieve intramammary pressure. Irrespective of prepartum milking, secretions obtained prior to calving are generally high in protein and low in lactose compared with normal milk but with relatively small concentrations of specific milk proteins. Because extensive removal of mammary secretions prior to calving can alter the course of mammary development (i.e., premature mammary cell differentiation), it is recommended that prepartum milking once initiated should continue until onset of normal milking at calving.

High concentrations of protein in secretions obtained prepartum in cows reflect the accumulation of immunoglobulins transferred into secretions from the bloodstream. As the alveolar cells become more differentiated and acquire the capacity to synthesize and secrete specific milk components, accumulation of immunoglobulins is reduced. However, the accumulated immunoglobulins in the secretions obtained with the first milking (suckling) postpartum (i.e., colostrum) provide passive immunity to the offspring. This is particularly important for those species that lack immunoglobulin transfer to the fetus in utero. With the onset of regular milking or suckling, the composition of mammary secretions progressively changes to reflect normal milk composition.

Across species, there are dramatic differences in milk composition. Milk from Holstein cows (the source of the majority of milk for human consumption in Western societies) has about 3.2 percent protein, 3.4 percent fat, and 4.6 percent lactose. In contrast, hooded seals produce milk with about 6 percent protein, 50

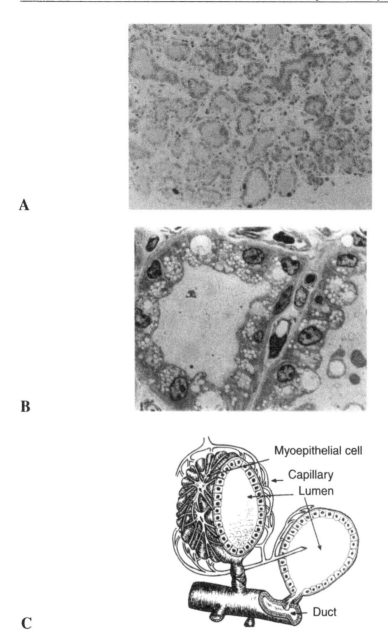

A

B

Myoepithelial cell

Capillary

Lumen

Duct

C

Fig. 1.2 **A** is an image from a histological section of mammary tissue prepared from the mammary gland of a pregnant ewe on day 115 of gestation (Smith et al., 1989). Note the presence of several spherical alveoli. Alveolar lumena are stained because of accumulated secretions. Small unstained areas contained lipid in the original specimen. Because the tissue sample was incubated with tritiated thymidine prior to fixation, this autoradiograph also shows a number of individual cell nuclei that were synthesizing DNA during the incubation period. **B** shows a portion of two alveoli from the mammary gland of a lactating cow. Note the presence of rounded basally displaced nuclei, abundant secretory vesicles, and larger lipid droplets. **C** illustrates the three-dimensional structure of an alveolus and associated duct. This figure is taken from Larson, 1985.

percent fat, and virtually no carbohydrates. Much of this variation likely reflects evolution-induced responses, which provide the best stratagem for offspring survival. Because the seal pups are born on potentially unstable pack ice, they must rapidly gain sufficient strength and insulation to survive. In fact, mothers suckle their pups for only 4 days during this period, but pups can double their 20 kg birth weight. While the hood seal has the shortest lactation of any known mammal, the high-fat milk provides the pup with the energy and metabolic water necessary for an abrupt introduction into a polar environment.

Evaluation of Mammary Development

Certainly, once lactation is initiated, milk production provides a measure of mammary development. For example, differences in milk production between beef and dairy cows are partially explained by the fact that lactating dairy cows have larger udders. However, it is also recognized that lactating cows with similar physical udder characteristics may produce very different amounts of milk. Ultimately, measures of mammary gland development and lactation performance must take into account both the quantity and functionality of the mammary parenchyma. Consequently, single measures of mammary function or structure rarely are adequate for accurate evaluation.

Tissue, Cells, and Organization

Since the mammary alveoli are responsible for milk synthesis and secretion, an accurate measure of the total number of mammary alveoli in the mammary gland would provide an excellent indicator of milk production capacity of the lactating animal. Assuming all of the secretory cells of each alveolus were equal both within and between animals, such a measure would certainly reflect lactation capacity. However, there is no device that can be used to make this determination. This means that the number of alveoli or secretory cells in the lactating mammary gland can only be estimated by indirect means. Just as important, most of these techniques give little indication of cell function. Furthermore, development of the mammary gland prior to lactation provides the tissue foundation for alveolar morphogenesis. Understanding mammary development and function also requires an appreciation of other tissues and cells within the mammary gland. The stromal tissue, which surrounds the milk-synthesizing alveoli, provides a framework for structural support and anchorage of parenchyma as well as spaces for passage of blood and lymphatic vessels. This combination of cells, tissues, and extracellular molecules is essential for the development of functional mammary parenchyma and the success of lactation.

Figure 1.3 illustrates the developing epithelial ducts in the mammary gland of a prepubertal ewe lamb. This low-power image demonstrates that as the epithelial ducts develop, they are surrounded by densely packed stromal cells and the stromal tissue is embedded in a sea of adipocytes. A higher-power view (Fig. 1.4) of a similar area from the mammary gland of a prepubertal heifer illustrates several

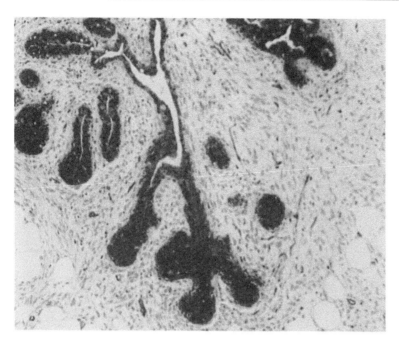

Fig. 1.3   A histological section of the parenchymal mammary tissue of a 4-week-old ewe lamb (×10). The developing ducts are darkly stained and closely surrounded by stromal cells. Masses of adipocytes and scattered bands of stromal tissue surround much of the outer boundary of the parenchymal tissue. Adapted from Ellis, 1998.

epithelial structures that radiate from a mammary duct. Pre- and postpuberty cross-sectioned ducts often demonstrate a scalloped appearance, suggesting a complex tubular structure. Indeed, recent three-dimensional computer animations prepared from serial sections of prepubertal bovine mammary parenchymal tissue elegantly confirm this tissue architecture. This suggests that in the ruminant, in contrast with the rodent, the gland during the prepubertal period is not filled with elongated ducts waiting for subsequent development of side branches. To use a plant analogy, in the peripubertal rodent, widely spaced mammary ducts fill the mammary fat pad like the bare branches of a tree. In the peripubertal ruminant closely packed ducts radiate from the gland cistern in broccoli-like fashion, but the ducts generally fill only a fraction of the mammary fat pad.

As the gland continues to develop, the relative area occupied by the epithelium increases at the expense of the surrounding stromal tissue. This is especially evident with the growth of alveoli during gestation. It is also evident in histological sections that much of the tissue area in late gestation and lactation is also occupied by lumenal space. This does not mean that there is necessarily a loss of stromal cells as the gland develops but rather that there is a dramatic rearrangement of cells and tissue elements so that that the stromal cells are less evident. As gestation advances, clusters of alveolar structures appear as scattered, round islands, until late in gestation histological fields are filled with closely packed alveoli (Fig. 1.2).

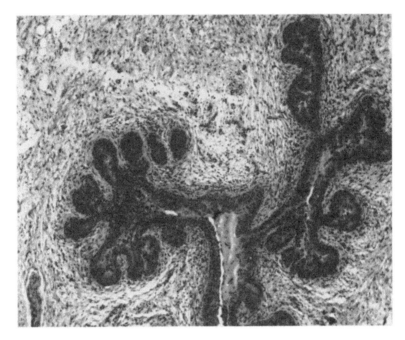

Fig. 1.4 This histological section of the mammary gland of a prepubertal heifer shows a number of epithelial structures radiating from subtending duct. H&E staining paraffin-embedded tissue (unpublished photomicrograph, Capuco et al., 2000).

Histological appearance of the rodent gland distinctly contrasts with that of the ruminant prior to gestation. Widely scattered ducts when cut in cross-section usually present a rounded profile indicating a simple tubular structure, and the entire structure is suspended in a sea of adipocytes interwoven with scattered blood vessels, fibroblasts, and leukocytes. Because the mammary glands of rodents are relatively flat, whole mounts of entire mammary glands can be easily prepared for examination. Such studies show that much of the ductular framework of the gland is established within the fat pad prior to the onset of puberty.

Prior to puberty, expansion of the mammary ducts into the fat pad depends on the activity of structures at the end of growth ducts called *end buds*. These club-shaped structures, 0.1 to 0.8 mm in diameter, occur at the elongating distal ends of the mammary ducts. From about 3 to 7 weeks of age in the mouse (mature virgin), end buds grow rapidly throughout the fat pad, occasionally bifurcating to form new growing ducts. Figure 1.5 shows part of a whole mount of a prepubertal mouse. The end buds are the apparent bulbous structures at the end of the darkly stained mammary ducts.

Indeed, it is believed that the end bud (denuded of extracellular matrix materials) can physically "push" between or around the adipocytes. This expansive growth continues until the end bud encounters the margin of the fat pad, at which time rapid growth ceases and the end bud regresses. It is also relevant that the individual end buds and ends of ducts avoid one another during this growth period. Consequently, with the onset of puberty, much of the growth involves appearance

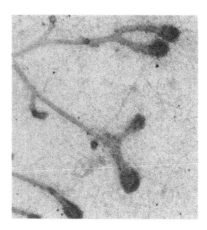

Fig. 1.5   Whole mount of the fourth inguinal mammary gland of prepubertal mouse. End buds are apparent at the endings of the ducts.

of side branches and development of alveolar buds. Figure 1.6 shows the whole mount of a mouse mammary gland during the prepubertal period. Notice that there are no end buds present and that the mammary area is occupied by widely spaced mammary ducts.

The preceding figures demonstrate the great utility of using rodents for mammary development studies. This is especially true for the study of mammary duct elongation and physiological studies for the period from birth into gestation. Results of treatments can be readily determined with observations and measurements of entire, preserved mammary glands. As an example, it possible to learn the precise effects of mitogens on specific mammary structures by implanting pellets with the test substance(s) directly adjacent to the structure in question (e.g., end buds).

How to Quantify Mammary Development

The examples given in the previous section confirm one of the primary problems with attempts to quantify mammary development. Although the focus is usually on the parenchymal portion of the gland, it is clear that changes in the nonparenchymal tissues of the gland are also dramatic. These changes do not necessarily occur in correspondence with parenchymal development. Consequently, techniques that fail to distinguish between these two primary tissue elements of the gland are often poor measures of mammary development. This is especially true for study during periods of marked mammary growth. In contrast, even simple measures of gland weight or gland volume can be of value during established lactation. Most measures can be divided into histological, biochemical, or physical techniques (Cowie and Tindal, 1971; Knight and Peaker, 1982).

Gross Measures

Palpation has been used to assess udder development in heifer calves and other ruminants. Attempts have not been satisfactory to make the technique semiquan-

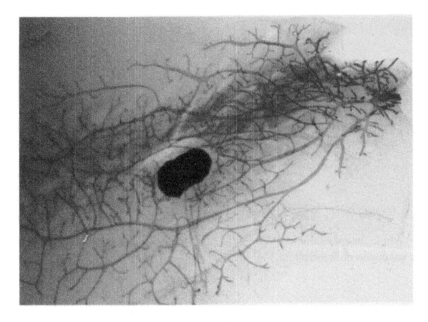

Fig. 1.6 Whole mount of mammary gland of a postpubertal mouse. The nipple area is to the upper right. Note the presence of mammary ducts radiating throughout the mammary gland but the absence of end buds.

titative with the use of calipers or measurement of teat spread and associated area calculations to estimate the extent of the mammary fat pad. Measurements are subject to large errors and are of limited value except to subjectively compare animals at similar physiological stages or to detect major problems with apparent mammary development in individual mammary glands of the udder. With some practice, it is possible to palpate the parenchymal tissue just above the teat in heifer calves or ewe lambs. This can be of value in some experimental situations but is not quantitative. Palpation of the udder of calves before about 1.5 months of age reveals only a small core of parenchymal tissue radiating away from the teat. Shortly thereafter, rapid growth of this parenchymal mass becomes evident, as a hard, dense, roughly walnut size and walnut-shaped mass. When removed for dissection, the teat and gland cisterns are apparent, and the outer portion of the mass is filled with highly compacted mammary ducts. Palpation of the udder is also useful in assessing mammary health related to milk removal and mastitis. By contrast, careful palpation of the breast in humans is an extremely important health care technique for early detection of breast cancer.

When animals are lactating, various studies have related measures of udder width, circumference, teat length, udder angle, width of rear teat attachment, and height to milk production or udder health. The correlation of these measures with performance or health in some studies is significant statistically but of limited biological value. Similar attempts to measure udder volume by water displacement or creation of plaster casts have been of value in limited experiments to relate changes in tissue volume with milk yield and udder storage capacity. Such studies have been most successful in goats because of the particular udder shape (relatively elongated and pendulous) and the ease with which volume measurements can be made.

Biochemical Indices

Realization in the early 1960s that the DNA content of cells is essentially constant (with the exception of the generally small proportion of cells that are undergoing DNA synthesis in preparation for cell division at a given moment) ushered in a host of studies to estimate mammary cell number based on total DNA content. This method is especially valuable when combined with careful dissection of the mammary gland to distinguish the parenchymal portion from the stromal tissue. Even with careful dissection of the mammary gland to remove apparent connective tissue, there are clearly nonglandular cellular elements (i.e., blood vessels, lymphatic vessels, nerves, fibroblasts, adipocytes, and white blood cells) that contribute to the DNA content of the parenchymal tissue compartment. Regardless, classic studies in a variety of lactating species give direct evidence that the number of mammary epithelial cells is proportional to milk production. Indeed, the correlation between total parenchymal DNA and milk production averages about 0.85. Consequently, any activity that reduces the number or function of the mammary alveoli will also reduce milk production. Other chemical methods developed to measure udder RNA and techniques to measure protein and fat have been adapted to the study mammary tissue (Tucker, 1981, 1987).

In addition to the nonsecreting epithelial cells and various stromal cells, extracellular secretions (i.e., proteins and proteoglycans that surround the mammary ducts and alveoli) are also critical for development. Collagen is the major extracellular protein component of the mammary stromal tissue. Moreover, the amino acid hydroxyproline is a specific and major component of collagen. Thus, assay of tissue content of hydroxyproline provides a quantitative measure of stromal tissue. When coupled with measurement of fat, the relative amounts of connective tissue associated with the parenchymal tissue can be estimated. However, during late gestation and lactation, the parenchymal fat content includes either lipids associated with milk synthesis (i.e., intracellular lipid droplets) or secreted lipids accumulated in lumenal spaces. Thus, no currently available chemical method provides a completely satisfactory means to distinguish the stromal elements incorporated within the parenchymal portion of the mammary gland. Data in Table 1.2 illustrate the dramatic changes in mammary growth from birth to lactation in Holstein heifers and crossbred ewes. Measured as trimmed udder weight or parenchymal DNA, mammary growth is greatest during gestation. However, relative lack of change in DNA from late gestation into lactation compared with trimmed udder weight suggests that DNA is a better measure of cell growth, since increased weight may be accumulated secretions.

Measurement of parenchymal tissue RNA and/or protein provides an indication of synthetic capacity and is most useful for evaluating the fully developed

Table 1.2 Mammary parenchymal growth in heifers and ewes

Measure	Stage of development				
	Prepuberty	Postpuberty	Midgestation	Late gestation	Lactation
Heifers					
DNA (g)	1.1	2.6	16.3	39.3	38.8
Wt. (g)	495	957	5,110	8,560	16,350
Ewes					
DNA (g)	0.02	0.09	1.3	2.3	2.6
Wt. (g)	15	78	557	1057	1340

Source. Data adapted from Smith et al., 1989; Keys et al., 1989; McFadden et al., 1990; and Sejrsen et al., 1982, 1986.

mammary gland. The reason for this is simple. As the secretory cells differentiate in concert with parturition, there is a marked increase in the presence of rough endoplasmic reticulum and corresponding production of mRNA for specific milk proteins. Consequently, onset of milk synthesis and secretion corresponds with a marked increase in the mammary tissue RNA. On a per cell basis, increased synthetic capacity can be evaluated by comparing the ratio of RNA/DNA in mammary parenchymal tissue. Late in gestation after alveoli have formed but before the onset of copious milk synthesis and secretion, this ratio is generally about one. During established lactation the RNA content of the secretory cells increases dramatically, and this ratio may be three or more. Changes in parenchymal tissue protein follow a similar pattern; however, care must be taken to account for the milk protein that may be trapped in the tissue. Finally, all of these measures are more useful if they can be applied on a whole mammary gland basis. For rodents or other animals, in which an entire mammary gland (or parenchymal tissue) can be isolated and sampled, these measures can be readily related to milk production. Measures of tissue composition derived from biopsy samples can be useful, but interpretation, especially if related to milk production capacity, must be made cautiously.

Histological Indices

The relevance of the microscope and histological preparations toward the study of the mammary gland is illustrated by the data given in Figures 1.1–1.7. Likely the simplest microscopic means of evaluating the mammary gland is the examination of thin slices of tissue with a dissecting microscope. Such slices are often defatted and stained and can yield much information about the structure of the mammary gland and mammary tissue even with the unaided eye. With the use of serial slices through the entire mammary gland, it is possible to estimate the area and with known thickness the volume of the mammary parenchyma and surrounding mammary fat pad. With the use of a dissecting microscope, it is possible to distinguish mammary ducts and clusters of developing lobules.

As demonstrated in Figure 1.6, the entire mammary glands of many rodents and other mammals can be evaluated in this manner because the glands are relatively thin and flat. Such examinations are most useful for glands obtained prior to lactation. Structures of the lactating gland are difficult to study even in rodents because the alveoli become tightly packed, the gland increases in thickness, and secretions and cells are densely stained. Nonetheless, even in glands from lactating animals it is possible to distinguish clusters of alveoli near the thinner edges of the gland. Figure 1.7 provides a series of images from murine mammary glands from the early postnatal period into lactation. During growth periods it is also possible to evaluate the area occupied by the mammary parenchymal tissue, but the pattern of mammary duct and alveolar bud growth is markedly more diffuse than in ruminants (Fig. 1.6). Because of this it is difficult to differentiate the area occupied by the epithelium and the adjacent stromal tissue. Consequently, it is actually easier to accurately and simply quantify the area or volume occupied by parenchymal tissue in the developing ruminant mammary gland than in rodents.

While it is not possible to prepare whole mounts of the mammary glands of cattle or sheep, slices or steaks prepared from the entire mammary gland can be useful to evaluate the relative development. In fact, for studies of tissue composition in ruminants, it is routine to first prepare sequential slices of the udder to allow the dissection of the parenchymal tissue from the mammary fat pad. Even in unstained slices the extent of parenchymal tissue development can be distinguished. For example, Figure 1.8 illustrates the growth of the mammary parenchyma of representative Holstein heifers treated with a placebo or recombinant bovine somatotropin (GH). Conformation of this effect on mammary parenchymal development was quantitatively confirmed by measurement of parenchymal DNA and computerized tomography (Sejrsen et al., 1986).

Technology Advances

As the comments above suggest, there has long been a need for the development of noninvasive methods to quantitatively evaluate development of the mammary gland and to specifically distinguish parenchymal and stromal tissue development. For human breast cancer detection, increased use of and technical advancements in mammography have been important in early detection of mammary lesions. Application of CT scanning to udders of slaughtered cattle or sheep has also been successfully used to measure amounts of parenchymal tissue and stromal tissue. The technique is sophisticated and capable of distinguishing the relative amounts of stromal tissue even within the parenchymal region of the gland. However, with respect to evaluation of treatment effects on mammary development, the method is really no better than the combined use of dissection and assay of parenchymal composition.

It is imaginable that modern nuclear magnetic resonance imaging (MRI) equipment might be capable of such studies in living animals, but the expense and logistics of scanning cattle or other large animals is prohibitive. Nonetheless, Fowler et al. (1990a) reported that MRI estimates of the mammary parenchymal volume of

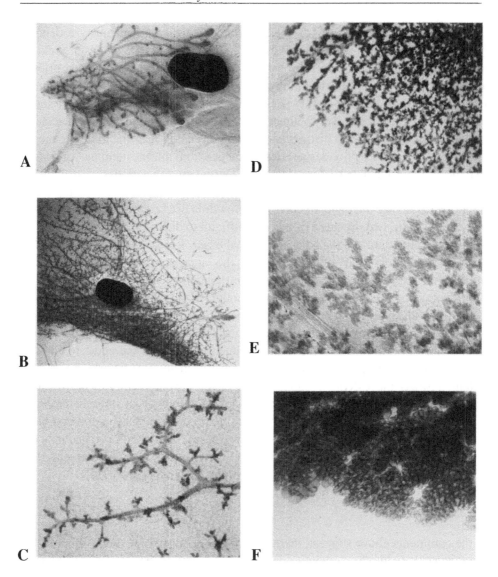

Fig. 1.7 Images depict a series of mammary whole mounts of the inguinal mammary gland of mice at various stages of development from the prepubertal period into lactation. **A** shows the inguinal mammary gland of a 4-week-old mouse. Ducts have grown from the nipple region (*left*) to just beyond the lymph node. End buds are also evident at the ends of several ducts. **B** shows a large portion of a whole mount of the mammary gland of an 8-week-old mouse. By this time ducts have elongated through the entire fat pad, and in areas alveolar buds and branches have appeared. **C** shows an area with formation of alveolar buds and branching in greater detail. **D** shows a small portion of the murine mammary gland on day 12 of gestation. Notice the now abundant appearance of alveolar structures. **E** is an enlarged image of the mammary gland at day 12 of gestation. Clusters of alveolar buds are clearly associated with a major duct. **F** shows the portion of a murine mammary gland on day 4 of lactation. Because of the massive development and accumulation of secretions, the rounded structure of clusters of lobules and alveoli is evident only at the thinner margins of the gland.

A

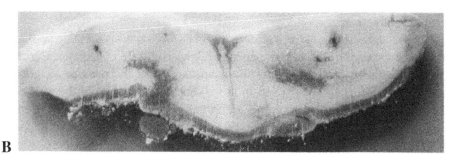

B

Fig. 1.8 Photograph of transverse sections of a frozen udder of a GH-treated (**A**) and placebo-treated (**B**) prepubertal Holstein heifer. The dark masses oriented above the teats of the individual glands indicate the presence of parenchymal tissue. Differences in parenchymal content of GH- compared with placebo-treated heifers were quantitatively confirmed by assay of dissected tissue mass, DNA content, and CT scanning. The figure is taken from Sejrsen et al., 1986.

goats were greater than those estimated postmortem but that the estimates were highly correlated. The technique yielded an excellent estimate of fluid volume in the udder (i.e., before and after milking). These workers were also able to quantify changes in parenchymal tissue volume during the course of gestation (1990b) and in response to hormonal induction of lactation (1991). During either natural or hormonal induction of lactation, the parenchymal volume was positively correlated with milk yield. Finally, specific changes in the MRI signal characteristics, so-called T_1 values, were seemingly related to expected changes in cellular composition of the parenchymal tissue during development. Specifically, the parenchymal T_1 value was lower in early gestation than late gestation for first-lactation goats and lower at the initiation of induced lactation. Because lower values are associated with fatty tissues, this suggests that changes in T_1 values reflect changes in tissue development, specifically the progressive replacement of adipocytes with epithelium. Perhaps advances in technology will lead to the creation of a new generation of portable, affordable scanning devices suitable for use in other animals.

Cytology Indices

Clearly, understanding of mammary development and production requires detailed measures of the total and proportion of the various cell types in the gland.

However, this information alone is of little use for estimating the functionality of the parenchymal tissue. It is possible that the parenchymal tissue contains many epithelial cells but that the cells have a limited capacity for synthesis or secretion of milk components. With histological studies of the mammary gland, it became apparent that parenchymal tissue from glands of high-yielding animals seemed to have an abundance of structurally highly differentiated polarized alveolar cells. Application of electron microscopy has demonstrated the similar appearance of the alveolar cells across species. The percentages of various cellular organelles within the cytoplasm of alveolar secretory cells from lactating rats and cows are illustrated in Table 1.3.

Despite the general similarity of lactating cells between species, these data quantify the appearance of only those cells with a well-differentiated appearance. Even a casual survey of mammary tissue from animals at the peak of lactation suggests there is substantial variation in the number of these well-differentiated cells both within and between alveoli. Careful analysis of mammary tissue at the light microscopy level, especially if the tissue is embedded in a plastic resin (compared with more traditional paraffin), allows an estimation of the proportion of alveolar cells, which fall into various classes of structural differentiation. This type of evaluation is especially useful for monitoring the dramatic changes that occur in the mammary epithelium at the onset of lactation or the effects of specific treatments that impact differentiation. Late in gestation, the alveolar cells are very poorly differentiated. The cells lack polarity and exhibit a large nuclear to cytoplasmic ratio, the presence of large lipid droplets, but few cytoplasmic vacuoles. Intermediate differentiation is characterized by a smaller nuclear to cytoplasmic ratio, the presence of a few apical vacuoles, and a medially to basally displaced nucleus. Fully differentiated cells display abundant supranuclear vacuoles, apical lipid droplets, basally located, spherical nuclei, and a large cytoplasmic to nuclear ratio.

Table 1.3 Ultrastructural analysis of cytoplasmic organelles in well-differentiated mammary epithelia from lactating rats and dairy cows

Cytological parameter	Rats	Cows
RER[a]	14.7	16.1
Golgi	20.9	18.8
Lipid	1.4	4.7
Mitochondria	7.1	6.3
Nucleus	21.7	22.0
Other	34.2	31.1

Source. Data adapted from Nickerson and Akers, 1984.
Note. Data are expressed as mean percentage cytoplasm occupied by each cytoplasmic component.
[a] Rough endoplasmic reticulum.

Table 1.4   Effect of colchicine on structural differentiation of the alveolar epithelium around the time of parturition

	Milk yield[a]	*Light microscopy*[b]			*Electron microscopy*[c]	
		Undiff. (%)	*Intermed. (%)*	*Full (%)*	*RER*[d] *(%)*	*Golgi (%)*
Control	60	7	12	81	18	25
Treated	17	49	46	6	7	9

Source. Data adapted from Nickerson and Akers, 1983, and Akers and Nickerson, 1983.
[a] Milk yield is given as kilograms produced per udder half during week 3 postpartum.
[b] Light microscopy data are the percentage of epithelial cells classified as undifferentiated, intermediately differentiated, or fully differentiated.
[c] Electron microscopy data are the percentage of cellular area occupied by rough endoplasmic reticulum or Golgi membranes and vacuoles.
[d] Rough endoplasmic reticulum.

Table 1.4 illustrates the effect of prepartum intramammary infusion of colchicine on subsequent milk yield and mammary epithelial cell differentiation in Holstein heifers. Two diagonal mammary glands of each of two pregnant Holstein heifers were infused with colchicine every second day from 1 week prior to parturition until calving. Twice daily milking began at calving, and the drug treatment was discontinued. Data for mammary biopsies obtained on day 21 of lactation are given. Clearly measures of mammary epithelial cell structural differentiation correlate well with function.

Summary

As is often the case, there are no easy solutions for quantification of mammary development. The best choice depends on the circumstances. However, likely the best advice is to include when possible a variety of measures that allow a full characterization of the parenchymal tissue (i.e., the organization and arrangement of the cells and if warranted the degree of cellular differentiation as well as determination of total parenchymal mass of the mammary gland). Essentially, combinations of anatomical and biochemical measures are more likely to yield satisfactory results than either alone.

Structural Aspects of Mammary Development

As the preceding sections illustrate, the mature mammary gland contains a variety of tissue and cell types. The mammary gland is also one of only a few tissues in mammals that can repeatedly undergo growth, functional differentiation, and regression. This is one of the reasons for the very great interest in the study of the mammary gland. The term *mammogenesis* refers to the development of mammary gland parenchymal structures. In usual circumstances, studies of mammogenesis usually are focused on the very large changes in the mammary gland, which begin around the time of puberty or more likely during pregnancy. However, the foundation for the dramatic mammogenesis during these stages begins when the animal is an early fetus. Fundamentally, mammary development can be developmentally considered a joining of the epithelial ectoderm and the underlying mesoderm.

Fetal Development

The first indication for the presence of mammary development is a slight thickening of the ventrolateral ectoderm in the embryo at the time limb buds begin to lengthen. This thickening is variously referred to as the mammary band, streak, or line. The details described here relate to the bovine. However, except for the timing related to gestation length, the sequence is similar across species.

Structural Precursors of Mammary Glands

The mammary band first appears at about 30 days in the bovine embryo. These condensed ectodermal cells appear on either side of the midline and extend from the upper limb to the lower limb. They are considered the first indicator of the cells destined to create the mammary gland. A further condensation of these cells is described as the mammary streak. By the time the embryo reaches the fifth week, the mammary streak becomes the mammary line. The mammary line is usually described as a thin ridge of ectodermal cells closely adjoining a deeper layer of compacted mesenchymal cells. As the mammary line shortens and more of the ectodermal cells proliferate and grow into the lower mesenchymal cell layer, the structure becomes the mammary crest. One relevant feature is that a distinct basement membrane continues to separate the growing ectodermal cells from the underlying mesenchymal cells. When shortening of the mammary crest is complete, the structure remains prominent in those areas where the mammary glands will finally form. For the bovine this is in the inguinal region.

As the ectodermal cells continue their inward growth into the mesenchymal layer, the appearance of a dome-like cluster of the cells at about day 40 can be defined as the mammary hillock. The further rounding of cells by day 43 of gestation and appearance of a hemispherical structure give rise to the mammary bud. In the bovine there are four mammary buds in the inguinal region, two on each side of the former mammary band. These buds ultimately will give rise to the fore and rear quarters of the udder. The mammary buds appear when the crown-rump length of the embryo is about 25 mm. In the human embryo the mammary bud is recognizable at about 20–30 mm or about day 49. Mammary bud formation is a critical stage in mammary development. Soon after the bud appears, the ectodermal layers sink into the mesenchyme, forming a depression (mammary pit) on the embryo's surface. The mammary bud stage marks the beginning of development patterns that distinguish mammary structures between various species. Differences between patterns of development in males and females also become apparent after this period.

Gland Number, Sexual Differentiation, and Species Variation

Up to the mammary bud stage, it is not possible to distinguish any differences in mammary development between males and females. Thereafter, female buds are more oval and have a smaller volume compared with males. It is also observed that the buds in the female appear closer to the surface than in males. Subsequently, early teat formation at about day 65 of gestation produces a more pointed teat in females compared with a somewhat flattened teat end in males. Thereafter, mammary development becomes slower and ultimately much less pronounced in males. Moreover, there is no nipple formation in male rats and mice; neither do male horses develop teats.

Bovine teat development begins with rapid growth of the mesenchyme surrounding the mammary bud so that the bud is pushed up toward the ventral surface. These events lead to an invasion of the mammary bud cells into the mesenchyme so that the mesenchymal cells are pushed aside. During this early development the mammary epithelium proliferates into relatively undifferentiated embryonic mesenchyme. Thereafter, blood vessels also begin to form in the mesenchymal area surrounding the epithelial bud. This probably is an indication of maturation of the mesenchyme. This elongated mass of epithelial cells evolves to become the primary sprout by about 80 days of gestation (crown-rump length 120 mm). Soon after the primary sprout gives rise to several secondary sprouts. The primary sprout leads to eventual formation of the teat and gland cistern, and the secondary sprouts to the major ducts, which drain into the gland cistern. Initially, the sprouts are solid cores of cells and are subsequently canalized. Precise mechanisms for canalization are not understood but are generally believed to involve both apoptosis and cell migration. As the teat develops, the tip invaginates from the outside, so that the stratified surface epithelium cells become keratinized, resulting in formation of the streak canal. This ultimately forms the keratin lining of the streak canal.

A dense mesenchyme with many fibroblasts continues to develop in close association with the mammary bud in those species examined. The mesenchyme is made up of precursors that give rise to the population of stromal cells that remain closely invested with the mammary ducts in the postnatal mammary gland. In the mouse the mesenchymal precursors of a mammary fat pad develop to the rear of the mammary bud on about day 14 of gestation. These cells differentiate into a well-defined fat pad, which consists mostly of adipocytes as well as necessary stromal elements (fibroblasts, nerves, and blood vessels). The first sign of a developing mammary fat pad appears at about day 80 in cows and about day 18 in rats at about the time when the secondary sprouts become evident (Sheffield, 1988).

From about the third to the fifth month of fetal development, the mammary bud of the human embryo changes very little. At this time the surface of the mammary bud appears to enlarge, and a depression forms at the surface. This is analogous to the appearance of the mammary pit in the bovine. Proliferation of the proximal region of the bud causes the appearance of 10 to 25 secondary buds. These are similar to the secondary sprouts of the bovine; they then gradually lengthen as solid cords of epithelial cells that penetrate the underlying mesenchyme. These structures are progressively canalized and form the major lactiferous ducts. These epithelial cords gradually branch at their ends. Each epithelial cord will correspond to a lactiferous duct and opening on the surface of the nipple. By contrast in the mouse embryo, only one cluster of epithelial cells grows from the mammary bud. This is called the primary mammary cord, and it gives rise to a single lactiferous sinus in the teat. Like the cow, the mouse has a single opening in each teat or nipple (see Table 1.1).

Although most species studied develop a mammary fat pad, this is not true in the human mammary gland. The adult female breast contains considerable amounts of adipose tissue, but it is not closely associated with the mammary ducts and does not appear until puberty. Normally, human mammary ducts grow and elongate in a much more fibrous stromal tissue than rodents'. In this respect the development of the ruminant mammary gland follows a similar path. Specifically, although there is adipose tissue in the developing gland, the developing ducts proliferate closely surrounded by a tunic of dense stromal tissue (see Figs. 1.3 and 1.4; Akers, 1991).

Summary

Given the widespread use of rodents as models for human mammary development and breast cancer, it is ironic that the developmental pattern for humans more closely follows that of ruminants. Regardless, it is clear that the pattern of mammary ductular development is regulated by the surrounding mesenchyme. Elegant tissue recombination studies have shown that mammary epithelial tissue grown with salivary mesenchyme develops into a glandular structure with a salivary pattern. However, the epithelial cells maintain a mammary pattern of biochemical differentiation since they secrete milk proteins when stimulated with lactogenic hormones. Secondly, androgen-induced regression of the mammary buds in male rodents is mediated by the surrounding mesenchyme. Thus, normal development of the embryonic mammary epithelium depends on specific stromal tissue requirements.

Birth to Conception

Generally little or no true lobulo-alveolar development occurs before conception. This period is associated with creation of a framework to allow proliferation of the secretory cells needed for lactation. The period is generally considered a time during which the duct system is extended and growth of the adipose and connective tissue increases. Ductular growth is limited for dogs, cats, and rabbits but more extensive for rats, mice, the rhesus monkey, cattle, and sheep.

Since it is expected that most organs in growing animals would also grow in concert with overall body growth, the regulated, cyclic development of the mammary gland can be difficult to study. One approach is to evaluate mammary development in terms of relative growth and to ask if growth of the mammary gland fits the law of simple allometry $y = bx^\alpha$. A usual approach is to log transform variables associated with body and tissue growth and use linear regression analysis to calculate the equilibrium constant (α), which relates the difference in growth rate of the organ under study to the growth of the body as a whole. Simple body weight or body weight$^{2/3}$ (to approximate surface area) is usually used as the independent variable in these analyses. The dependent variable (mammary gland area, mammary gland weight, DNA content, weight of parenchymal mass) serves as an index of the growth of the mammary gland. When $\alpha = 1$, growth is said to be isometric. If $\alpha > 1$, growth is said to be positively allometric (simple allometry). If $\alpha < 1$, then growth of the organ is negatively allometric (enantiometry).

Data for rodents, cattle, and sheep demonstrate that from birth until before puberty the mammary gland grows isometrically with the rest of the body. Thereafter, the rate of mammary gland growth becomes allometric usually through the first few estrous cycles. Growth then reverts to a period of isometric growth that also waxes and wanes during the course of the estrous cycle. As demonstrated for heifers in the classic study by Sinha and Tucker (1969), positive allometric growth begins at about 3 months of age and continues until about 9 months of age. Specifically, when mammary development was expressed as the log of mammary DNA and body growth as a log of body weight, the equilibrium constant (α) averaged 1.6 in the period from birth to 2 months, 3.5 from 3 to 9 months, and 1.5 between 10 and 12 months of age (Fig. 1.9A). These data indicate that the mammary gland grows somewhat faster than the body during the early postpartum period but dramatically faster between 3 and 9 months of age.

Although the absolute amount of mammary growth during the period prior to conception is only a fraction of the mature mammary gland growth late in gestation (see Table 1.2), there is compelling evidence that alterations in mammary growth during the peripubertal period can dramatically affect subsequent mammary function. For example, rapid prepubertal weight gain inhibits subsequent mammary parenchyma tissue growth and reduces subsequent milk production (Sejrsen and Purup, 1997). Not surprisingly, very few cattle studies have considered changes in mammary development in the period shortly after parturition. Our observations of mammary glands of Holstein calves between 1 and 3 months of age, while minor in terms of absolute mass, suggest that parenchymal tissue per gland increases approximately 20-fold in only a few weeks. It is intriguing to consider that growth during even this very early postnatal period may also be critical to the success of lactation.

It can be assumed that mammary growth and development are markedly stimulated by the onset of ovarian activity at puberty. Prepubertal ovariectomy dramatically inhibits mammary growth in many species. Prepubertal development of the mammary gland of mice, rats, and heifers is 3- to 10-fold greater in intact compared with ovariectomized animals. Interestingly, prepubertal mammary development of ewes is not impacted by ovariectomy. Table 1.5 presents a relative

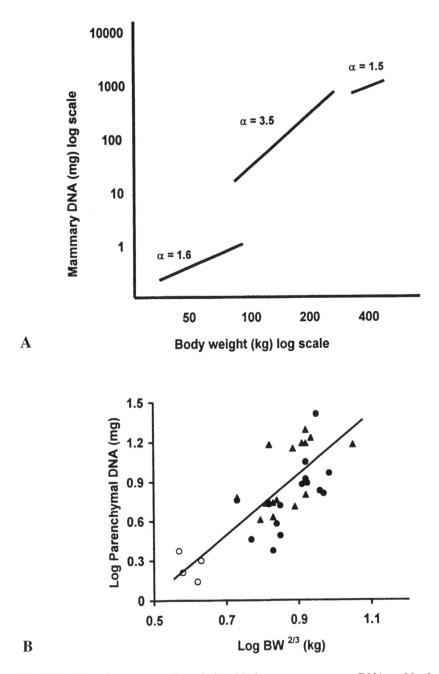

Fig. 1.9 Chart **A** represents the relationship between mammary DNA and body weight in Holstein heifers. Segments illustrate periods: birth to 2, 3 to 9, and 10 to 12 months of age. Redrawn from Sinha and Tucker, 1969. Chart **B** gives a relative growth analysis of ovine mammary parenchymal development in prepubertal ewes. Log-transformed variables were used in regression analyses to calculate the allometry coefficient for intact (▲) and ovariectomized (●) lambs. The open circle is for starting animals. The solid line represents the regression equation. Ovariectomy had no significant effect on the observed allometry coefficients (P > 0.05). Chart **B** is taken from Ellis et al., 1998.

Table 1.5 Estimates of relative mammary gland growth in intact and ovariectomized prepubertal ewes

Mammary development measure	Corresponding α value (Body wt.$^{2/3}$)
Log parenchymal weight	3.10 ± 0.2
Log parenchymal DNA	2.37 ± 0.3
Log total gland weight	1.42 ± 0.1
Log total gland DNA	1.20 ± 0.2

Source. Data adapted from Ellis et al., 1998.

growth analysis of mammary development in ovariectomized and intact ewe lambs between birth and 12 weeks of age. These animals had their ovaries removed at 10 days of age. In contrast with rodents and heifers, ovariectomy did not prevent the onset of allometric mammary growth. Specifically, values for α were greater than 1 for measures related to mammary tissue growth generally (total gland weight or total gland DNA) but markedly higher for measures specific to parenchymal tissue. Neither was there a difference in magnitude of α between 2 and 6 weeks or 6 and 13 weeks of age. Thus, consistent positive allometric parenchymal growth was observed in both intact and ovariectomized animals (Fig. 1.9B). These results reinforce the idea that despite our general understanding of factors that regulate mammary development, there are likely many examples for which dogma does not apply.

Despite the fact that there is very little mammary tissue present at birth, those mammary cells that are present are capable of proliferation and a degree of mammary-specific cytodifferentiation. Indeed, the endocrine changes at parturition are believed to be responsible for precocious development and fluid secretion ("witch's milk") sometimes observed in human infants.

In an interesting twist, use of transgenic cows, pigs, rabbits, goats, sheep and mice for synthesis and secretion of therapeutic and nutritional proteins in their milk has fueled interest in premature induction of lactation, especially in transgenic cattle and sheep. This is because one of the limitations to producing transgenic cows for this purpose is that it often takes 2 or more years from the time a calf is born until its first natural lactation. A recent study (Ball et al., 2000) showed that lactation could be hormonally induced in prepubertal heifers and that composition of secretions was similar to that of normal milk. Neither did the induction treatment impair normal reproduction or subsequent lactation. These results suggested that hormonal induction of lactation in prepubertal heifers could be used as a technique for testing the fidelity and relative expression of recombinant proteins designed for secretion into milk prior to the first natural lactation. Thus, mammary glands of even sexually immature mammals are capable of proliferation and differentiation to initiate milk secretion.

Although normally there is little or no genetic selection of dairy cattle until initial production records become available at first calving, early selection could be practiced if methods existed to accurately predict subsequent milk production. Early attempts to equate udder development (based on palpation and recording of a score for development) with subsequent milk production showed that ratings were positively correlated but correlations were too low to be practical. Similarly, correlations between udder DNA and RNA at 5 months of age and subsequent milk production are positive (0.21 to 0.27) but low. Of course early selection of genetically superior sires could have even a greater impact than early selection of heifers.

It was recently shown (Filep and Akers, 2000) that it is possible to induce the secretion of casein in mammary tissue explants prepared from sexually mature bulls if the tissues are incubated with lactogenic hormones. Comparisons of tissue from selection and control line bulls showed that tissue from high-genetic merit bulls have a greater capacity for casein secretion. Thus, specific physiological measurements applied to the mammary tissue of potential sires might well provide information useful in sire selection.

Conception to Parturition

Under normal circumstances, true alveoli are not formed until conception. Variation among species is almost certainly related to the number of estrous cycles that occur prior to conception. Using cattle as an example, in the early stages of pregnancy, the duct system continues to develop with appearance of a rudimentary lobulo-alveolar system by about 5 months of gestation.

Parenchymal and Stromal Tissue Development

It is estimated that 94 percent of mammary development for the hamster takes place during gestation. Estimates for other species range from 78 percent for the mouse and sheep to 66 percent in rabbits and 60 percent for rats. By far the greatest promoter of natural mammary growth or mammogenesis is pregnancy and its associated hormonal and physiological changes. With the influence of pregnancy, mammary growth is reinitiated after reversion to isometric growth following puberty. This growth can be described by an exponential equation with the following form: $Y = a^{bt}$, where Y = mammary size, t = day of gestation, and a and b are constants. Such equations have been developed to model mammary growth in cows, goats, and guinea pigs. The variable a is mammary gland mass or size at the beginning of gestation and the variable b is the first order rate constant. The time necessary for the mammary gland to double is equal to $ln2/b$ (Sheffield, 1988).

Measurement of mammary DNA or weight of dissected parenchymal tissue is useful to quantify rates of mammary growth, but these techniques do not distinguish differences among cells types. Problems of trying to distinguish between stromal and parenchymal tissue is especially difficult in rodents. For example, both wet weight and defatted dry weight of the inguinal mammary glands of rats are reported to vary from 50 to 120 percent during pregnancy, but total mammary DNA changes 200 to 300 percent. Since total DNA is a reflection of cell number, changes in weight alone underestimate cellular development. Table 1.6 illustrates changes in mammary development of ewes during gestation. Changes in weight, total DNA, or percentage epithelium illustrate a marked increase in parenchymal tissue between day 80 and 115 of gestation. However, between day 115 and 140, quantitative histology (percentage epithelial area) alone suggests there is no continued parenchymal development. This is clearly not the case since both tissue weight and total DNA doubles. Lack of change in epithelial area between day 115 and 140 reflects that alveoli are present and at these stages of gestation, but an increase in lumen and decrease in stromal tissue area reflect accumulation of some secretions and compression of stromal tissue between alveoli. Lack of difference in alveolar cell number in cross-section also suggests alveoli are relatively uniform between periods. In addition to the appearance of alveoli during gestation, the cells undergo a progressive structural and biochemical maturation as parturition approaches. The increase in labeling index on day 115 reflects the rapid growth during this period.

Appearance of Alveoli and Secretory Cells

Figure 1.10 shows changes in the growth of mammary parenchymal of beef and dairy heifers during pregnancy and into lactation. Comparisons between Charts A and B illustrate that measures of udder parenchymal mass alone can be misleading. Specifically, based on udder parenchymal mass, there is marked growth of parenchymal tissue between day 260 and day 49 of lactation in both Holsteins and Herefords (Fig. 1.10A). In contrast, when udder growth is evaluated based on

Table 1.6 Effect of gestation on mammary growth, histology, and epithelial cell labeling index in ewes

	Day of gestation			
Measurement	Day 50	Day 80	Day 115	Day 140
Trimmed udder weight (g)	304 ± 43	253 ± 30	557 ± 29	1050 ± 188
Total DNA (mg)	94 ± 4	57 ± 14	1304 + 152	2324 ± 321
Epithelium (%)	14.2 ± 0.3	19.2 ± 2.1	41.2 ± 2.3	40.3 ± 1.8
Stroma (%)	85.8 ± 0.3	77.8 ± 3.2	35.7 ± 6.5	20.7 ± 3.0
Lumen (%)	NP	1.0 ± 0.2	23.1 ± 4.3	39.0 ± 2.9
Alveolar cell no.	NP	NP	36.4 ± 3.8	36.6 ± 1.4
Labeling index	0.16 ± 0.03	1.0 ± 0.2	3.6 ± 1.7	1.0 ± 0.4

Source. Data adapted from Smith et al., 1987.
Note. NP indicates not present, specifically alveoli appeared between day 80 and 115 of gestation. Labeling index indicates the percent of epithelial cells that incorporated tritiated thymidine following a one-hour incubation of tissue explants at the time of slaughter.

parenchymal DNA (Fig. 10B), marked differences between breeds remain, but little change occurs between day 260 of gestation and day 49 of lactation in Holsteins. This is likely explained by the increased accumulation of secretions in the Holstein heifers compared with the Hereford heifers. When lactation performance is considered, it is clear that differences in milk production (3.5 vs. 20.3 kg/day) are explained by changes in udder parenchymal tissue mass and function. For example, total RNA as well as the RNA/DNA ratio is greater in Holstein than Hereford heifers. Interestingly, the ability of mammary explants from lactating animals to secrete α-lactalbumin (a specific milk protein) in culture closely mirrors (57 vs. 289 ng per mg of tissue per 24 hours) the corresponding 5.8-fold difference in daily milk production.

After these data were published, we subsequently completed a detailed cellular evaluation of mammary parenchymal samples collected from these heifers. We determined percentage tissue area occupied by epithelium, stroma, and lumenal space and the number of cells per alveolar cross-section. There was a small but consistent increase in average number of cells per alveolar cross-section in the dairy heifers during gestation and a statistically significant increase in the lactating Holstein heifers. Differences were small, but the impact of alveolar diameter on alveolar volume is substantial. If alveoli are assumed to be spherical, alveolar volume ($V = 4/3\ \pi r^3$) is easily calculated. If radius of an alveolus is 50 μm, volume is 522,023 μm^3. An increase of only 10 μm (approximate width of an epithelial cell) nearly doubles average alveolar volume (902,059 μm^3). Changes in alveolar volume and consequently storage or secretory capacity are likely important in selection for increased milk production. Indeed, the difference in average number of cells per alveolus during lactation between breeds (36 vs. 49) suggests a 2.6-fold greater average alveolar volume for Holstein heifers. Of course, these interpretations assume that comparisons between beef and dairy heifers would mirror the effects of selection responses of dairy heifers. Milk yield is clearly a function of the number of secretory cells and their activity. Simplistically, milk yield is directly impacted by factors that increase or maintain the number of alveoli available during lactation as well as factors that influence functional differentiation of these cells.

Mammary Cell Differentiation

Comparisons between breeds of cattle also indicate the significance of cytological differentiation of alveolar cells to explain differences in milk yield. For lactating Holstein and Hereford heifers, alveolar cells were characterized as poorly differentiated, intermediately differentiated, or fully differentiated as illustrated and described previously (see the section "Cytology Indices"). More than 40 percent of the alveolar cells in Herefords were poorly differentiated, but in Holstein heifers nearly all of the cells were classified as intermediate or fully differentiated. These results provide additional evidence to support the idea that much of the difference in milk production between beef and dairy cows depends on increased mammary

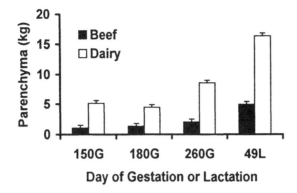

A   **Day of Gestation or Lactation**

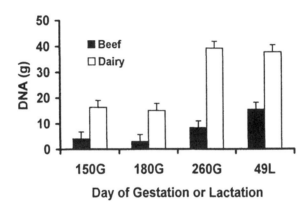

B   **Day of Gestation or Lactation**

Fig. 1.10   Chart **A** illustrates growth of the mammary parenchyma of beef and dairy heifers during pregnancy and lactation based on weights of parenchymal tissue dissected from udder slices obtained at slaughter. Chart **B** provides a similar data, except growth is indicated from a measure of total parenchymal tissue DNA. Both measures indicate a consistent advantage for Holstein compared with Hereford heifers. However, the increased weight of parenchymal mass between day 260 prepartum and day 49 of lactation for Holsteins is not observed when growth is measured based on DNA content. The data illustrate the importance of applying multiple measures when evaluating mammary development and likely indicate a marked increase in cell growth in beef but not dairy heifers late in gestation and into early lactation. Figure adapted from Keys et al., 1989.

function. Indeed, calculations indicate 22 percent and 55 percent of the 5.8-fold greater milk production in lactating Holstein heifers in the study were due to differences in cell mass and function, respectively.

After the initial formation of lobulo-alveolar tissue during the second half of gestation, individual alveoli continue to increase in size, and new alveoli are added until most of the mammary area is filled with alveoli. In fact for most species, the general histological appearance of parenchymal tissue is very uniform with abundant profiles of sectioned alveoli, large areas of alveolar lumen, and compressed stromal tissue between alveoli from late gestation into lactation. Relative area occupied by epithelium and lumenal space are similar, ~40 percent each with the remaining ~20 percent occupied by stromal tissue. This, however, does not mean that the tissue is quiescent with the approach of parturition. The alveolar cells progressively undergo the biochemical and ultrastructural differentiation necessary for the onset of copious milk secretion (lactogenesis) at the time of parturition.

Although it has long been thought that terminally differentiated cells do not proliferate, several observations suggest that preparation for the onset of milk secretion does not prevent proliferation. Figure 1.11 provides photographs of sectioned mammary tissue from a sheep on day 140 of gestation, from a cow about 2 weeks before calving, and from a cow 2 weeks after calving. In panel A portions of each of three alveoli from the pregnant ewe are shown. It is apparent that a number of the cells have accumulated large lipid droplets and that spaces of the alveolar lumina are darkly stained. The staining is an indication of the accumulation of protein (likely primarily immunoglobulins associated with colostrum formation). The autoradiographic image also shows each of three alveolar cells that have incorporated tritiated thymidine into their nuclei. The density of the labeling indicates that the incorporation is associated with DNA synthesis rather than DNA repair. Thus, at least initial stages of lactogenesis are compatible with continued cell proliferation of these cells. Panel B shows a photograph of mammary tissue 2 weeks before calving. Alveolar lumina are very darkly stained. A number of lipid droplets are present both within the cells and lumenal spaces. Secretions have accumulated so that the alveolar cells are compressed. Panel C illustrates the highly differentiated alveolar cells common with the onset of lactation. Because the tissue was obtained shortly after milking, the cells are seemingly taller than in panel B, but more significantly the cells have acquired the structural features of actively secreting cells. A lacy appearance highlights the apical region of the cell because of the abundance of secretory vesicles in contrast with the darkly stained basal-lateral cytoplasm (ergastoplasm). The fully differentiated cell is polarized with the basolateral area devoted to the uptake of precursors and synthesis of proteins and lipids and the apical cytoplasm, with now abundant Golgi, devoted to posttranslational modification of proteins and packaging of proteins and lactose for secretion from the cell.

Lactating Mammary Gland

Mammary growth as measured by total DNA continues during the early days of lactation in rats, rabbits, guinea pigs, and mice. In the rat for example, as much a 26 percent of the total mammary growth occurs during lactation. In the mouse there is a transient surge in mammary cell proliferation 2 to 3 days postpartum. In fact it is reported that the cell population doubles between the last day of gestation and day 5 of lactation. The guinea pig is especially interesting in that there is

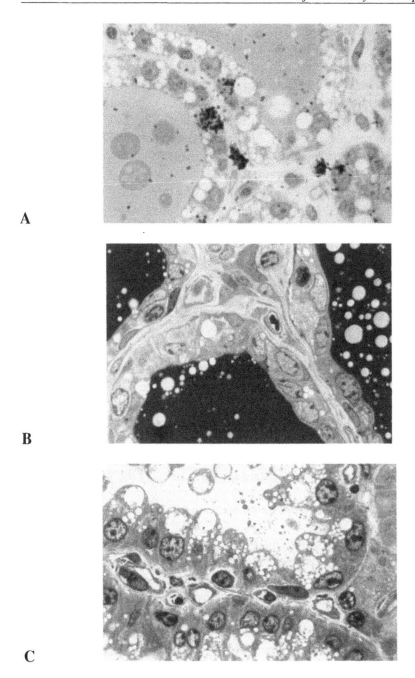

A

B

C

Fig. 1.11 Photomicrographs of mammary tissue taken from pregnant sheep and lactating cows. **A** is an autoradiogram of mammary tissue from a ewe on day 140 of gestation. The dark grains indicate ^{3}H-thymidine incorporation into cell nuclei. Note there are three labeled alveolar cell nuclei despite the presence of limited lumenal secretions and intracellular lipid droplets. **B** a cow 2 weeks before calving shows portions of three alveoli with darkly stained luminal contents and poorly differentiated cells. **C** is of mammary tissue from a cow 2 weeks after calving. Note the appearance of well-differentiated alveolar cells.

little change in total mammary DNA during gestation but a dramatic increase within 2 days of the birth of the pups. However, if suckling is not permitted in these species, the increased growth just after parturition is prevented. This suggests that signals associated with suckling or milking are important for growth especially in early lactation. Regardless, it seems clear that once lactation is established that the rate of mammary cell proliferation is markedly lower than during other stages of mammary development. However, recent reports show that several weeks of "extra" milking stimulation within the first 2 months of lactation in dairy cows increases subsequent milk production when milking frequency is returned to normal. Mechanisms for the effect are unknown, but the fact that it continues after treatment supports the idea that it may involve recruitment of additional well-differentiated cells. This might reflect overt proliferation of additional secretory cells or perhaps enhanced functional differentiation in a population of pre-existing non-secretory cells (Bar-Peled et al., 1995). This effect contrasts with the galactopoietic treatment of cows with bovine somatotropin (bST) in that milk production is dramatically increased during the period of treatment but returns to control levels when treatment is discontinued. This pattern of response has been interpreted to indicate that bST alters metabolism rather than cell proliferation. This is clearly an area of intense research interest.

In dairy species it is usually concluded that there is normally little mammary growth during established lactation. However, comprehensive data for early lactation are lacking. The concentration of DNA in mammary parenchyma is relatively unchanged in early lactation in sheep, goats, or cows, but it is risky to interpret this as a lack of mammary growth since the entire mammary gland was not evaluated. Total parenchymal DNA doubles between 2 weeks before and after parturition (27.9 vs. 46 g), but these data do not determine if the growth occurs before or after parturition. Rates of thymidine incorporation are very low for mammary tissue taken from lactating ruminants, but mitotic cells are sometimes observed.

In lactating goats unilateral inhibition of secretion in one gland results in a compensatory increase in milk production of the other gland. There are also reports of increased milk production in uninfected quarters of cows with mastitis. The relative contributions of hypertrophy and hyperplasia for this effect are unknown. However, in lactating beef cows covering one half of the udder to prevent sucking increased thymidine incorporation in lactating mammary glands compared with control glands of cows with continued suckling of all glands. Morphologically, lactating tissue from control and compensatory treatment cows was indistinguishable, and the tissue did not appear consistently different between zones within lactating quarters. About 40 percent of the parenchymal tissue consisted of closely packed alveoli. In these areas the secretory cells appeared highly differentiated similar to cells from dairy cows. Remaining parenchymal tissue contained alveoli more widely scattered in the stromal matrix, and the cells were less well differentiated. Labeled cells were observed in both regions but appeared more frequently in the less well-differentiated tissue (Akers et al., 1990).

Onset of Lactation to Mammary Involution

Once initiated, lactation depends on regular suckling or milking of the mammary gland. Although the time required for regression varies markedly between species (i.e., days for rodents vs. weeks for ruminants), without milk removal the alveolar structure is eventually degraded, alveolar cells dedifferentiate, and many cells undergo apoptosis. Without the stimulus of another gestation, the gland progressively reverts to a structure similar to that of the mature virgin. However, in dairy cows milk production increases with each successive lactation. This suggests accumulative mammary growth with each lactation cycle. Normal husbandry dictates that dairy cows are rebred soon after the onset of lactation and milked for much of the concurrent pregnancy. Consequently, during the later part of lactation, the cow has dual functions of growth of the developing calf and continued lactation. Compared with wild ruminants or beef cows, the period between consecutive lactations is relatively short. This means there is less opportunity for mammary regression or involution. Secondly, hormonal and metabolic changes associated with late gestation and preparation for parturition are conducive for mammary development. This means that the time course of mammary involution is impacted by concurrent gestation.

The Lactation Curve

Lactation curves for typical Holstein cows, a record Holstein producer, and a human for comparison are illustrated in Figure 1.12. However, it is important to realize that estimates of milk yield need to be evaluated in relationship to composition of the secretions obtained, state of maturity of the offspring, suckling patterns, as well as diet and behavior of the species. Practically, production curves are most reliable for dairy animals (cows, goats, sheep) because these animals have been selected for ease of milking and handling, high yields of milk, and their response to machine or hand milking.

For other species direct measurements of milk yield require that the animals are often sedated and/or injected with exogenous oxytocin to obtain milk. Under these circumstances the relevance of the data (other than comparison between treatments) can be questioned. Certainly, it is unknown if yields reflect "natural" levels of production or if measures of composition are accurate. An alternate for estimating yields is the response to timed-suckling episodes. This typically involves removal of the suckling young for a period of time, then weighing the litter or offspring immediately before and after a set period when returned to the mother (Beal et al., 1990). For species in which the offspring are solely dependent on milk for nourishment, measurement of offspring weight gain can also be used as an indirect measure of milk production of the mother.

While there is likely some question of the accuracy of yield estimates for nondairy mammals, comparisons of peak milk yields between species give valuable insight into physiological demands of lactation. In particular, scaling milk production relative to body weight or mammary gland weight suggests common relationships and illustrates the effect of genetic selection for milk production. Although only a small number of the known mammals have been evaluated, when data for 19 species including mouse, hamster, rabbit, pig, human, goat, sheep, cow, and camel are used, unselected species array in a regular pattern for relationships between milk yield and body weight, energy output in milk and body weight, and mammary weight versus body weight. Equations for these relationships include the following examples (Hanwell and Peaker, 1977):

(1) Milk yield (kg/day) = $0.084 \times$ body weight $(kg)^{0.77}$
(2) Energy output in milk (kcal/day) = $127.2 \times$ body weight$^{0.69}$
(3) Mammary weight (kg) = $0.045 \times$ body weight $(kg)^{0.82}$

Per unit of body weight smaller animals have higher milk yields, greater outputs of energy in their milk, and relatively larger mammary glands than physically larger animals. If data are averaged, the relationship between mammary gland weight and milk yield can be expressed by the following equation: milk yield (kg/day) = $1.67 \times$ gland weight $(kg)^{0.95}$. Because the exponent does not differ significantly from unity, on average, 1 g of mammary tissue from a lactating animal produces 1.67 ml of milk per day. There must certainly be some variation in this ratio (i.e., variation in the degree of differentiation of the alveolar cells, contribution of accumulated secretions to weight measurements, stage of lactation, milking or suckling patterns), but the value provides a reasonable rule of thumb for comparative purposes. It comes as no surprise to good dairy cattle managers that energy and metabolic demands on the high-producing dairy are very great. The energy and substrate demands to support increased milk production are met by a combination of increased feed intake and mobilization of body reserves. Indeed, the typical dairy cow is in a negative energy balance for much of early lactation. However, when compared with smaller mammals, these demands seem relatively less severe. This is a reflection of not only the demands of the suckling young but also the fact that smaller mammals typically secrete a much more energy-dense milk than cows. For example, as an extreme case it is estimated that the pygmy shrew (*Sorex minutus*), which weighs only 5 g, must more than double its food consumption during lactation and eat more that four times its body weight per day to meet the additional demands of lactation. Lactation in small animals generally involves a relatively short period of intense metabolic demand but in larger animals a longer period of lesser demand.

The relationship between milk yield and body weight is illustrated in Figure 1.13. The regression line for several species not selected for milk production is shown. Because certain species have been selected for high yields (dairy cows, dairy goats), values for some of these selected animals are also plotted as individual data points. Data points for these animals are above the calculated regression line for the nonselected species (i.e., well above 95 percent confidence intervals), reflecting greater than expected milk production. The success of genetic selection for milk yield in dairy cows, a trait with a heritability of only 0.2

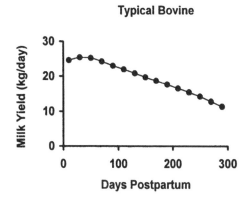

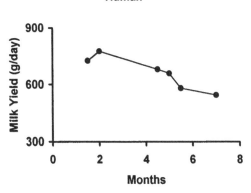

Fig. 1.12 Lactation curves for typical lactating Holstein cows (**A**), a Holstein production record holder (**B**), and a human (**C**). Data adapted and from Mepham, 1983, and Grossman et al., 1999.

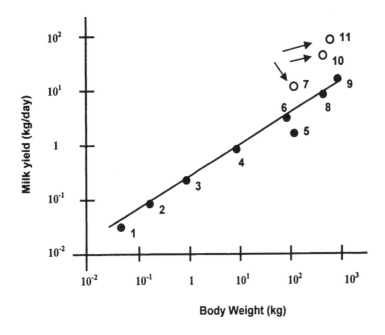

Fig. 1.13 Relationship between peak milk yield and body weight in dairy (open circles) and nondairy mammals (filled circles). 1 = mouse, 2 = hamster, 3 = rat, 4 = fox, 5 = human, 6 = goat, 7 = dairy goat, 8 = sheep, 9 = beef cow, 10 = Holstein cow, 11 = Jersey cow. Adapted and redrawn from Hanwell and Peaker, 1977.

to 0.3, is such that a typical Holstein cow has been converted to the metabolic size of a dog and exceptional producers to the metabolic size of rodents. Concern for good husbandry practices and health of dairy animals particularly in face of recent techniques (e.g., use of bovine somatotropin) to increase production and continuing genetic progress is justified, but it is equally important to view these physiological demands on the dairy cow within an appropriate frame of reference (Mepham, 1983).

Mammary Involution

Mammary involution can be stimulated at any stage of lactation by removal of suckling young or in dairy animals by suspending milking. Most of the detailed work on mammary involution has focused on acute induction of involution. However, after the peak of lactation, a gradual involution occurs as the young are progressively weaned. In dairy cows there is also a progressive decline in milk production with time even with regular milking. This decline is greater for cows that become pregnant soon after calving (i.e., less persistency of lactation), but milk yield slowly declines irrespective of a concurrent gestation.

Details of mammary involution have been studied in greatest detail in rodents. Regardless of the differences in timing of events between species (i.e., days to hours in rodents vs. days to weeks in ruminants), involution involves apoptosis (or programmed cell death) of alveolar epithelial cells and tissue remodeling. Apoptosis of mammary epithelial cells occurs in several phases of mammary development not just mammary involution. These include canalization of major mammary ducts in fetal development, rapid elongation of mammary ducts via the terminal end bud, and the waxing and waning of ductular growth during estrous or menstrual cycles. Apoptotic destruction of abnormal cells at early phases of carcinogenesis is also a likely mechanism for protection against breast cancer.

In rodents removal of the suckling young and accumulation of milk at midlactation rapidly initiates involution of the mammary gland. Marked changes in gene expression are evident within 24 hours, and evidence of widespread apoptosis observed by 48 hours. For example, in rats, expression of mRNA for the caseins is reduced 95 percent and that of acetyl-CoA carboxylase, a key lipogenic enzyme, 98 percent within 24 hours. Translational actively is also reduced. A second stage of involution is initiated between 72 and 96 hours with the activation of a series of tissue proteinases including stromelysin-1 and -2, gelatinase A, and plasminogen activator. By this time remodeling of the mammary gland to prepare for a new reproductive cycle is well underway. Given the rapidity and magnitude of these tissue changes, it is not surprising that involution in the mouse is only partially reversible if suckling young are returned within 48 hours. Although much of the involution process is initiated by milk stasis, the continued secretion of lactogenic hormones associated with milking or suckling can delay the process. This is shown by experiments in which suckling is allowed in only selected glands. For example, in lactating sheep when both glands were nonsuckled for the first 15 days postpartum, mammary prolactin receptor was reduced 84 percent, but in the nonsuckled gland of ewes with the opposite gland having continuing suckling stimulation, the prolactin receptor was reduced only 36 percent. Parenchymal DNA concentration was not affected by suckling treatments; however, RNA concentration followed a pattern similar to prolactin receptor, lowest when neither gland was suckled (2.1 mg/g), highest in suckled glands (7.4 mg/g), and intermediate in nonsuckled glands companion to a suckled gland (3.8 mg/g). Alveolar structure was maintained in all treatments, but the cytological appearance of the epithelial cells reflected changes in RNA concentration. Epithelial cells from ewes with neither gland suckled were poorly maintained with many cells engorged with secretory vesicles and lipid droplets.

Involution in the ruminant mammary gland is decidedly slower with less loss of alveolar structure. After three days of nonmilking, casein and α-lactalbumin mRNA were reduced, but β-lactoglobulin mRNA was unchanged. After a week α-$_{s1}$casein and α-lactalbumin mRNA were dramatically lower (85 and 99 percent respectively). After 4 days of nonsuckling, alveolar structure is degenerated in the rodent, and apoptosis, based on the degree of DNA laddering, is nearly maximal. However, even in nonpregnant, lactating beef cows, alveolar structure was largely intact in the absence of suckling with only isolated areas of tissue degeneration evident even after several weeks. Nonetheless, based on quantitative histology, well-differentiated cells, common in suckled glands, were rare in nonsuckled glands. Interestingly, regression of the gland in the absence of sucking was not uniform. The structure of parenchymal tissue distant from the teat was better maintained. Even 42 days after cessation of suckling, areas of alveolar structure were present. Overall, alveoli from lactating glands had more cells per cross-section (30.4 ± 0.9) compared with 21.4 ± 0.8 in glands not suckled for 42 days. In total these data illustrate that even a prolonged period of nonsuckling (42 days) does not result in complete destruction of the mammary alveoli. However, since total gland DNA was reduced (50 to 64 percent after 21 and 42 days of nonsuckling, respectively), cells are lost with prolonged milk stasis in the beef cow. Regardless, survival of dedifferentiated alveolar cells should allow milk secretion to be reinitiated if milking is resumed. In dairy cows resumption of milking after 12 days of nonmilking in selected glands caused a rapid recovery in milk production to near pretreatment values for these glands. When none of the glands were milked, recovery was impaired (i.e., only 28 percent of pretreatment yields were obtained). In contrast, return of suckling calves to their dams after a 28-day hiatus in beef cows was recently shown to restore milk production to about 50 percent of control. After 1 week of renewed suckling, milk yield had increased 11-fold (2.2 kg/d), and composition was near normal except for reduced lactose (3.7 vs. 4.7 percent). These data do not mean that mammary involution does not occur in cattle; however, the extent and timing of events are more prolonged than for rodents (Akers et al., 1990; Lamb et al., 1999).

When extrapolated to the dairy cow with a dry period between lactations occurring 40-60 days prior to the birth of the calf, it is reasonable to suggest that tissue degeneration is certainly less than in rodents and likely less than in nonpregnant cows. Just as suckling or milking of contralateral glands acts to slow involution in companion nonmilked or nonsuckled glands, the involution process is also likely minimized during the typical dry period of dairy cows. Certainly, during the later stages of the dry period, the mammary gland is growing, and alveolar cells are preparing for lactogenesis. This overlap of lactation and pregnancy when cows are "dried off" in preparation for calving means those stimuli associated with mammary involution and milk stasis conflict with mammogenic and lactogenic stimuli of pregnancy. Cows are usually well into the last trimester of pregnancy (i.e., 40 to 60 days before calving at drying off), goats may be in early stages to second trimester of pregnancy, and ewes are typically nonpregnant at the end of lactation. Differences in the management of the reproductive/lactation of these dairy animals are due to the seasonal nature of reproduction in goats and sheep, gestational length, and usual lactation periods. Among dairy animals the dairy cow is unique in this abrupt interface between successive lactations.

Dairy managers have developed methods to maximize profitability and milk production. Empirical evidence shows that the inclusion of a nonlactating or dry period between lactations is needed to maximize milk production in the dairy cow. Without a dry period milk production in the subsequent lactation is reduced about 20 percent. Thus, the dry period is not essential but is beneficial. In fact a dry period of 40–60 days seems optimal, since a dry period that is too short (less than 40 days) also impairs subsequent milk production. The dry period seems to be important because of effects on the mammary gland rather than for effects on the nutritional status of the dam as was once commonly believed.

Because of the typical breeding pattern, little attention has been given to determining an optimal dry period for goats or sheep. However, when lactating goats were hormonally induced to ovulate and mate during the usual seasonal anestrous period, they entered the next lactation without a dry period, and they produced 12 percent less milk than controls. However, the data were confounded with possible effects of season. Consequently, in a second study these researchers used a within-animal experimental design to determine the effect of drying off on lactation in goats. One gland was milked continuously, and the other dried off 24 weeks before parturition. There was no difference in subsequent milk production between udder halves. In contrast with the first study, this suggests a dry period is of no benefit. However, the dry period was relatively very long (i.e., three times the optimal length for the cow). Finally, the udder half experimental design may be less than ideal for this evaluation. Interactions of glands of differing lactational states within the udder on cell turnover and lactogenesis have not been investigated. Since prepartum milking advances lactogenesis and milk production, milking one gland may conceivably advance lactogenesis and milk production in other glands. Certainly, in nonpregnant ruminants milking or suckling delays mammary involution in companion nonsuckled glands as indicated above. Thus, it is possible that milking one gland during the prepartum period inhibits the ability of the opposite gland to produce maximally during the subsequent lactation or that milk production was increased in glands milked continuously when the opposite gland was dried off. The importance of a dry period to maximize milk production in dairy goats and sheep is unsettled. Studies to determine the effects of pregnancy status, length of the dry period, and other implications of the udder half experimental design are warranted (Wilde et al., 1999).

Unique Aspects of Involution in the Dairy Cow.   Similar to the response to acute milk stasis, mammary regression throughout the dry period in dairy cows is markedly less than in other species. Neither sloughing of epithelial cells into alveolar lumina nor detachment of cells from the basement membrane is observed. As indicated previously, accurate assessment of changes in mammary cell number requires a combined evaluation of tissue morphology and total udder DNA.

Toward this end, multiparous lactating cows and dry cows were slaughtered at 53, 35, 20, and 7 days prepartum corresponding to 7, 25, 40, and 53 days into the dry period. During this 60-day dry period, there was no evidence that a net loss of mammary cells (involution or regression) had occurred in the nonlactating cows. At no time did the mammary glands from nonlactating cows contain less total DNA or parenchymal mass than in lactating cows (Fig. 1.14A). Morphometric analysis of mammary tissue samples showed that tissue area occupied by epithelium did not decline and that alveolar structures remained intact during the dry period.

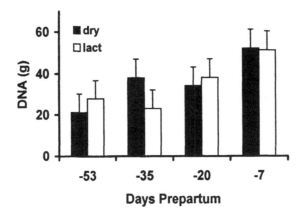

A

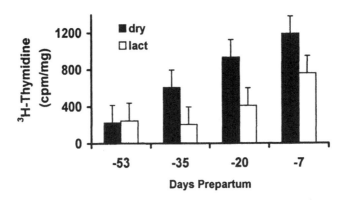

B

Fig. 1.14 Changes in mammary glands of cows milked continuously or dried off 60 days prior to expected parturition. **A** illustrates changes in total parenchymal DNA. DNA content was not significantly affected by treatment. However, total DNA was approximately doubled by 7 days prepartum in both groups. A pattern of increased DNA synthesis in dry cows compared with milked cows across the prepartum period suggests a greater turnover of cells in cows given the usual dry period treatment (**B**). Redrawn from Capuco et al., 1997.

Nonlactating and lactating cows entered the dry period with equal numbers of mammary cells based on total parenchymal DNA. However, calculation of cell types in tissue sections showed that the proportion of epithelial cells in nonlactating cows (83 percent) was greater than in lactating cows (74 percent) 1 week before parturition. Rate of ^{3}H-thymidine incorporation was consistently greater in mammary tissue from nonlactating cows after 1 week into the dry period (Fig. 1.14B). Autoradiography of tissue sections showed that this increased incorporation was due to an increase in the percentage of alveolar cells incorporating the nucleotide.

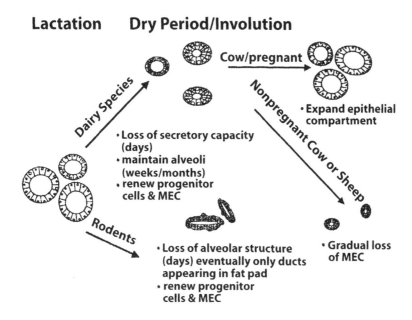

Fig. 1.15   Comparison of structural changes during mammary involution in dairy animals and rodents. Note the protective effect of drying off with a concurrent pregnancy in dairy cows. MEC = mammary epithelial cells. Redrawn from Capuco and Akers, 1999.

Since the total cell number was not different between lactating and nonlactating cows during the dry period, the increased DNA synthesis likely indicates increased cell turnover in cows given a dry period. This supports the idea that the dry period is important because changes in this interval of time enhance the replacement of senescent alveolar cells prior to the next lactation (Capuco et al., 1997).

To summarize, milk production is a function of the number of secretory cells and the secretory activity per cell. The concept that old or senescent alveolar cells need to be replaced to maximize production is based on two related ideas. The first is that old cells do not secrete as efficiently or for as long as new cells; the second that old cells have reduced proliferative capacity so that the number of secretory cells decline prematurely during lactation. If these ideas are correct, the persistency of lactation in cows without a dry period should be less than for cows given a dry period. This hypothesis while logical is difficult to test. In the absence of sufficient cell renewal during lactation, milk production would decline because of progressive loss of alveolar cells and perhaps decreased production per cell. Comparisons between breeds suggest that both cell number and function per cell explain reduced milk production in nondairy breeds. Experiments with goats suggest that the decline in milk production during lactation is due to a progressive loss of secretory cells but not activity per cell. In cows it appears that both loss of cells and reduced activity per cell explain the loss in persistency of lactation. Cells that need to be replaced during the dry period may be those that are responsible for expanding and maintaining the number of secretory cells (Capuco and Akers, 1999).

Recent studies in mice (Chepko and Smith, 1999) indicate that the mammary gland contains a population of stem cells that are capable of reproducing all of the differentiated cells in the mammary gland, as well as sets of progenitor cells with a more limited capacity to generate either ductular or alveolar cells. This subset of progenitor cells has limited replication potential and is renewed from the primordial stem cell population. Moreover, this subset of cells undergoes apoptosis during mammary involution, so they evidently must be replaced during the non-lactating period to maximize milk production. If stem cells exist in the bovine mammary gland, the dry period is undoubtedly important in the renewal of this population (Fig. 1.15).

Chapter 2
Mammary Development, Anatomy, and Physiology

Overview of Interactions—Focus on the Dairy Cow

The success of milk synthesis and secretion depends on coordinated activity of all the physiological systems of the body. At times the demands of mammary development or lactation test the physiological capacity of the mother, but long-term maintenance of homeostasis is essential for the survival of both the mother and offspring. Interestingly, studies testing the long-recognized positive effect of bST to increase milk production led to the realization that at least part of the physiological basis for the increased production of genetically superior cows involved differences in partitioning of both dietary and tissue-derived nutrients. Related to homeostasis, the effects of bST to promote increased milk production came to be described as homeorrhetic. *Homeorrhesis* was coined as a term to describe physiological adjustments, essentially chronic coordination of physiological processes and tissue metabolism, to support a particular physiological activity. In the context of milk secretion in high-producing dairy cows, this refers to bST (or other factors) that mediate physiological changes to increase mobilization of tissue nutrients, absorption of dietary nutrients, changed blood flow, and increased feed consumption to meet the requirements for increased milk production.

Blood Supply

Angiogenesis

Despite the obvious importance of having an ample blood supply to support milk production, study of blood vessel formation has received little attention in normal mammary development and almost none in dairy animals. In contrast, study of factors that stimulate vessel formation in breast tumors is an area of research focus critically important in human health. *Angiogenesis* refers to the formation of capillaries from pre-existing vessels in the embryo or adult organism.

45

Vasculogenesis concerns the development of new blood vessels from the differentiation of endothelial cells precursors called *angioblasts* in situ. The advent of molecular biology and the corresponding capacity to identify the myriad of bioactive proteins and peptides in mammary tissue and in circulation have generated an impressive listing of growth factors and proteins capable of affecting either rates of endothelial cell proliferation in vivo or the degree of vessel formation in animal cancer studies.

For example, vascular endothelial growth factor (VEGF) and its family of related proteins are believed to be major regulators of vasculogenesis and angiogenesis in normal tissue development as well as pathological conditions. VEGF has a host of effects on the endothelial cells. These include the ability to maintain viability, stimulate mitogenesis and chemotaxis, and change permeability. Effects of the VEGF are mediated by two or more tyrosine kinase receptors. VEGFR-1 and VEGFR-2 are believed to be especially important, but other structurally similar receptors such as VEDFR-3 can also bind VEGF as well as the variants VEGF-C and VEGF-D, which are more important in formation of lymphatic vessels. The signaling pathways for these receptors are only partially understood, but it is generally recognized that VEGF is a dominant regulator of tumor angiogenesis and likely normal vascular development. For example, the mammogenic growth factors insulin-like growth factor I (IGF-I) and epidermal growth factor (EGF) are known to stimulate the tissue expression of VEGF. Other growth factors involved include endothelins, fibroblast growth factors, and transforming growth factors. It is important to recognize that there are multiple isoforms for each of these growth factors. Lastly, analogous to development of mammary ducts, elongation and proliferation of growing vessels into the stromal matrix of the mammary gland must involve complex interactions between the endothelial cells of the vessel, extracellular matrix molecules, and specific stromal cells (Hansen and Bissell, 2000).

Aside from effects of growth factors and hormones, modulation of the extracellular tissue matrix components via extracellular proteases is certainly critical. It is apparent that mammary duct growth and angiogenesis require dramatic reorganization of the surrounding stromal matrix. Based on their mechanisms of action, four major classes of enzymes include serine, cysteine (thiol), aspartic (acidic), and metalloproteineases. In mammary tissue the metalloproteineases (stromelysins [1–3], gelatinases [A, B]) and the serine proteinase plasmin have received particular attention (Twining, 1994; Bernaud et al., 1998). One of the most striking examples has involved production of transgenic animals in which stromelysin-1 production is induced in the mammary tissue via linkage of the gene to a mammary-specific gene promoter. In these animals, the mammary glands exhibit precocious lobulo-alveolar morphogenesis and appearance of milk proteins by approximately 10 weeks after birth in virgin animals (Sympson et al., 1994). These data illustrate the dramatic impact that alteration of the extracellular matrix surrounding the mammary ducts or vessels can have on development. Other data suggest that plasmin is also likely important in mammary tissue remodeling. Briefly, inactive plasminogen occurs in virtually all body fluids and thus serves as a reservoir for proteolytic activity. Activation leads to formation of the serine protease, plasmin, which can degrade a host of extracellular proteins. The cascade of reactions leading to activation is in turn controlled by complex interactions between plasminogen activators and inhibitors. Indeed, high concentrations of u-PA (activator of plasminogen) in breast tumors are associated with a poor survival prognosis (i.e., rapid tumor growth and angiogenesis). In normal murine mammary development, both enzyme content and mRNA for u-PA were elevated in physiological states associated with rapid growth but reduced during lactation. In support of this in cultured proliferating bovine mammary cells, IGF-I increases secretion of u-PA as well as steady state levels of u-PA mRNA. These data suggest two general themes: (1) that the effects of IGF-I on the plasminogen system depend on the state of differentiation of the mammary gland and (2) that PA is elevated in concert with mammary development and angiogenesis (Politis, 1996).

In dairy animals it is intuitive to predict that hormones and growth factors, which affect mammary ductular proliferation or lobulo-alveolar formation, would also impact angiogenesis. Because the focus of most mammary developmental studies is on the epithelium, the stromal tissue and particularly blood vessel formation is often ignored. There is substantial variation in ^{3}H-thymidine labeling of endothelial cells in mature tissues from 0.01 percent to 0.14 percent for retinal and myocardial tissues, respectively. This is markedly lower than for many epithelial cells (i.e., labeling rates of 10 to 15 percent are common for intestinal epithelium, and even in the growth "quiescent" lactating mammary gland, epithelial cell labeling rates average about 1 percent) (Fig. 1.11). In the fully developed mammary gland a low rate of endothelial cell labeling is likely a reflection of only an occasional need for cell replacement or DNA repair within a stable vascular network.

In contrast, as the mammary gland develops, a vascular network must be created in conjunction with glandular tissue formation. Administration of exogenous estradiol to prepubertal heifers stimulated a 34-fold increase in the ^{3}H-thymidine labeling index of endothelial cells within 48 hours (0.16 vs. 5.5 percent), and the effect was specific since progesterone administration was ineffective. Given that estrogen can mediate the synthesis or secretion of a number of local tissue growth factors, it is unclear if its effect on endothelial cell proliferation is direct. Since there are marked changes in angiogenesis in the ovary, placenta, and endometrium during the normal reproductive cycle, it is logical to predict that steroid hormones, especially estrogens, are involved. In support of this idea, culture media conditioned by large preovulatory estrogen follicles stimulated the growth of bovine endothelial cells more effectively than media conditioned by estrogen inactive follicles. Since the mammary gland is ultimately a reproductive organ, a relationship between estrogen and angiogenesis during periods of stimulated mammary growth is logical but unproven.

Circulatory Route

The cow with its four inguinal mammary glands in the udder is supplied with arterial blood by branches of the posterior aorta and the internal and external iliac arteries (Fig. 2.1). The internal iliacs progress caudally, but the vessels supply

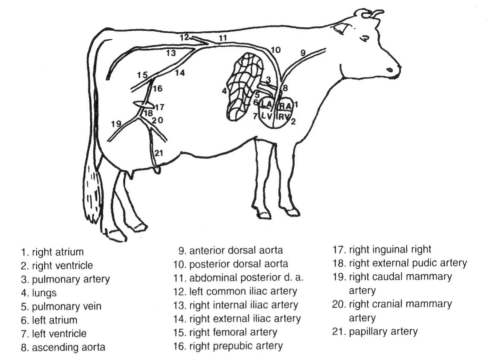

1. right atrium	9. anterior dorsal aorta	17. right inguinal right
2. right ventricle	10. posterior dorsal aorta	18. right external pudic artery
3. pulmonary artery	11. abdominal posterior d. a.	19. right caudal mammary
4. lungs	12. left common iliac artery	artery
5. pulmonary vein	13. right internal iliac artery	20. right cranial mammary
6. left atrium	14. right external iliac artery	artery
7. left ventricle	15. right femoral artery	21. papillary artery
8. ascending aorta	16. right prepubic artery	

Fig. 2.1 Diagram showing the arterial blood supply to the udder of the cow from the right atrium to the papillary artery of the teat.

blood primarily to the reproductive tract. However, paired branches, the perineal arteries, supply small amounts of blood to the udder. While the amount of blood flow is relatively unimportant for nutrient supply, since these vessels pass near the rear of the udder, they are frequently used as a source of mammary arterial blood in metabolism studies. The external iliacs proceed caudally and laterally to the pubis to become the femoral arteries. The femoral arteries give rise to the paired prepubic arteries, which pass to the internal inguinal ring where these arteries divide into the posterior abdominal and external pudic arteries. The external pudic arteries (also called *external pudendal arteries*) leave the abdominal cavity through the inguinal canal and enter the right and left halves of the udder. Once the vessels enter the mammary gland, they become the mammary arteries where branches supply blood to the large supramammary lymph nodes. The main mammary arteries further branch into cranial and caudal mammary arteries, which spread ventrally to supply the front and rear quarters on each side.

Blood draining from the udder leaves by two primary routes (Fig. 2.2). The first is the external pudic vein, which derives from the combination of the cranial and caudal branches of the mammary vein. Blood in this path passes to the right atrium of the heart in parallel with the corresponding arteries: prepubic, femoral, external iliac, common iliac, and posterior vena cava. The second is the subcutaneous abdominal or milk vein, which emerges from the anterior basal border of

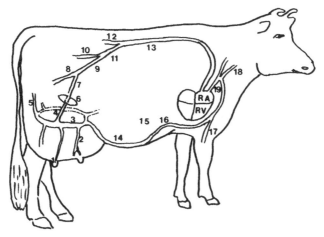

1. papillary vein
2. mammary vein
3. venous circle
4. right external pudic vein
5. right perineal vein
6. right inguinal ring

7. right prepubic vein
8. right femoral vein
9. right external iliac vein
10. right internal iliac vein
11. right common liliac vein
12. left common iliac vein
13. posterior vena cava

14. right subcutaneous
   abdominal vein
15. diaphragm
16. right internal thoracic vein
17. right brachial vein
18. right jugular vein
19. anterior vena cava

Fig. 2.2 Routes for venous blood from the mammary glands of the cow.

the udder as a continuation of the anterior or cranial mammary vein. The vessel enters the abdominal wall at the xiphoid process to unite with the internal thoracic veins. Blood subsequently enters the anterior vena cava. The perineal veins are also present, but the arrangement of the values is such that little blood actually leaves the udder in this manner.

Although the subcutaneous abdominal veins are impressive looking, movement of blood via this path is physiologically minor, since these vessels can be ligated with little or no effect on milk production. In virgin goats, sheep, and cows the external pudic vein carries essentially all the venous blood back to the heart. In multiparous animals the valves of the subcutaneous abdominal veins and perineal veins progressively fail so that the perineal vein drains blood from the vulva region where it is mixed with external pudic vein blood to travel back to the heart. This is important in metabolic studies because the perineal veins must be occluded when sampling mammary venous blood from the external pudic to ensure that the sample represents mammary venous blood. Blood sampled from the subcutaneous abdominal vein may be unrepresentative because of mixing and backflow.

The Lymphatic System

The lymphatic system is the final portion of the cardiovascular system to consider. Lymphatic vessels collect the interstitial fluids that bathe the cells. These vessels are structurally similar to the capillaries and veins and are scattered throughout the mammary gland. The vessels drain in a dorsal and caudal direction to progressively larger trunks to reach the large paired supramammary lymph nodes located in the mammary fat pad above the rear quarters of the udder. Like veins the lymphatic vessels have one-way valves to allow passive flow in the direction of the lymph nodes. Because the rate of flow can be slow, it is not uncommon for the udder to become edematous around parturition. One possible cause is accumulation of mammary secretions and associated increased pressure acting to partially occlude stromal tissue between alveoli to minimize passive flow. Second, prior to parturition the tight junctions (zona occludens) between the alveolar epithelial cells are not completely competent. This allows products sequestered in the alveolar spaces to more readily enter interstitial spaces. Presence of these materials in interstitial fluid and lymph acts to alter osmolarity and hydrostatic pressure so that less fluid is removed. An accumulation of interstitial fluid leads to swelling of connective tissue spaces and thus udder edema. A common treatment is massaging the udder in the direction toward the supramammary lymph nodes to increase the rate of fluid removal. After parturition tight junctions become structurally and functionally mature so that passage of milk components into interstitial space is minimized and the milk–blood barrier is maintained. Onset of frequent milking also acts to reduce pressure buildup. These events act to eliminate udder edema.

Once lymph gets to the supramammary lymph nodes, larger lymph vessels transport the fluid through the inguinal ring and via branches toward the deep inguinal, external iliac, or prefemoral lymph nodes. These paired nodes connect with a single lumbar lymphatic trunk. This large vessel progresses cranially to form an enlargement called *cisterna chyli,* which also collects lymph draining from the intestinal tract. The vessel continues to become the thoracic lymph duct, which empties into the anterior vena cava.

Significance of Blood Flow in Milk Production

While a detailed discussion of cardiovascular physiology is beyond the scope of this chapter, consideration of blood flow statistics serves to emphasize importance relative to milk production. Blood flow to the mammary gland supplies all of the nutrients necessary for milk synthesis and serves to carry metabolic waste products away. With a heart rate of 70 beats per minute, a 500 kg cow has daily blood cardiac output or volume of about 71,000 L. The lactating mammary gland uses about 10 percent of cardiac output even after considering cardiac output increases about 45 percent in late pregnancy and into early lactation. It is estimated that the ratio between blood flow and milk production is 500:1. Assuming 1 L of milk and 1 L of blood equal 1 kg, this means that a cow with a daily production of 35 kg of milk requires a flow of 17,500 L of blood through the udder. Clearly, excellent cardiovascular function is critical for the success of lactation.

Although overall mammary blood flow (MBF) and milk yield are positively correlated, MBF through the capillary beds surrounding the alveoli is really the essential feature of blood delivery important in milk production. However, regulation of flow through this microvasculature is poorly understood. Capillaries tend to form interweaving connections called *capillary beds*. The flow of blood from an arteriole to a venule is called *microcirculation*. These beds consist of essentially four components. Blood is supplied by the terminal arteriole, which delivers blood into a metarteriole. The metarteriole is continuous with a thoroughfare channel, which connects with a postcapillary venule and to larger veins. The true capillaries branch from the metarteriole segment of the structure. However, at these branch points smooth muscle cell sphincters surround the capillaries. If these cells are completely contracted, blood entering the metarteriole continues directly into the thoroughfare channel and exits via the postcapillary venule without ever entering the true capillaries. Changes in the degree of contraction of these sphincters are the essence of local control of blood flow. Since the contraction of muscle cells requires ATP, and formation of ATP is dependent on O_2, one of the simplest explanations for rhythmic local blood flow directly relates to O_2 availability. With a high pO_2, contraction occurs, and blood flow slows; with slower flow, less O_2 means reduced contraction (i.e., less ATP and thus vasodilation). Increased flow allows a new round of ATP synthesis, and the cycle repeats.

Certainly, there are acute local changes in MBF, and the presence of arteriovenous thoroughfares makes understanding and control of blood flow in local capillary beds difficult and complex. As in other portions of the body, mammary arterioles are supplied by sympathetic nerve fibers, but there is no evidence of parasympathetic fibers. As would be expected, stimulation of the sympathetic nervous system or administration of norepinephrine reduces mammary blood flow. Conversely, cutting the inguinal nerve causes vasodilation in the udder of sheep and cows. As is the case in other tissues, this suggests that innervation of the mammary gland acts to maintain the mammary vascular sphincters in a tonic state of constriction.

Regardless, local intrinsic control of mammary blood flow is certainly as important. Best described for blood flow in muscle tissue, classic local mediators of blood flow include O_2, CO_2, adenosine, lactic acid, and pH. In support of this idea, high-arterial pO_2 reduced MBF in mammary glands perfused in vitro. Several additional compounds have satisfied experimental criteria as putative metabolic regulators of local MBF. These include parathyroid hormone-related protein (PTHrP), IGF-I endothelin, and nitric oxide (NOX). PTHrP was initially described for having a role of humoral hypercalcemia in malignancy, but the protein was subsequently identified in milk and mRNA for the protein isolated from mammary tissue. In goats, the increase in overall MFP at parturition is preceded by an increase in PTHrP. Close arterial infusion of exogenous PTHrP also increases overall mammary MBF but not necessarily milk production. IGF-I is locally produced in the mammary tissue, and like PTHrP, infusion of this growth factor increases MBF (Prosser et al., 1996).

NOX has been the subject of much recent investigation. Originally called *endothelium-derived relaxing factor*, it is a potent vasodilator derived from the endothelium in the form of a free radical gas generated by oxidation of the amino acid L-arginine by the enzyme nitric oxide synthase (NOS). Its production is stimulated by increased secretion of the growth factor endothelin. Three isozymes of NOS include NOS-I, present in the brain and adrenergic nerve fibers; NOS-II, which is in macrophages; and NOS-III, primarily present in the vascular endothelium. At physiological pH, NOX exists for only a few seconds so that its action is likely restricted to very small areas near the site of synthesis. It is believed to rapidly diffuse to the smooth muscle cells of the arterioles to cause localized vasodilation via interaction with receptors that stimulate a guanylate cyclase–mediated second messenger signaling pathway. Infusion of diethylamine nonoate, a NOX donor, into the mammary gland in lactating goats increases MBF, but infusion of the NOS inhibitor, N^{ω}-nitroarginine, decreased MBF. Further support for a role for NOX in local MBF comes from histochemical studies of mammary tissue of goats and cows, which show antibodies against NOS-III localized in the endothelium and secretory cells of the alveoli. The presence of NOS within the mammary glands suggests that the alveolar cells may control their own blood and nutrient supply. Until recently it was generally assumed that the sympathetic nervous system ruled blood vessel diameter, but it is increasingly evident that its primary role is to maintain a degree of vasoconstriction that is modified by local release of vasoactive substances. NOX seems to be an especially important regulator in this regard. Interestingly, excess induction of NOX in macrophages, which home to the mammary gland in response to infusion of endotoxin as a model for mastitis, may be partially responsible for tissue damage associated with naturally occurring mastitis (Lacasse et al., 1996; Bouchard et al., 1999).

Nervous System

While development of the mammary gland and milk synthesis and secretion have only a minimal requirement for nervous system input (i.e., maintenance of homeostasis), neural control of milk removal is substantial. There is no direct evidence that nerve fibers directly affect the secretory activity of the alveoli. For example, isolated perfused mammary glands are capable of continued milk production for a considerable period of time. In the animal, completely denervated mammary glands continued to produce milk. This does not mean that the nervous system has no role in lactation. However, to appreciate the interactions between the nervous and endocrine systems to effect milk ejection, an understanding of the rudimentary organization of the nervous system is useful.

Organizational Overview

The neuron is the functional cell type of the nervous system because of its ability to transmit impulses between neurons or between neurons and other effector cells. The classic image of the neuron is the motor neuron. Short processes, the dendrites, serve to gather short-lived graded electrical potentials that, if they occur sufficiently rapidly or from simultaneous multiple events, allow the affected neuron to depolarize sufficiently for induction of an action potential. The all or none action potential induces a wave of electrical depolarization across the surface of the entire neuron and ultimately down the axon of the neuron. As the end of the axon terminal becomes depolarized as a result of an acute, transient influx of sodium, calcium is released into the cytoplasm of the nerve ending. Increased free calcium allows vacuoles of neurotransmitter stored in the nerve terminals to undergo exocytosis and release their contents to the outside of the cell. The anatomy of the nerve tracts is arranged so that release of the neurotransmitter occurs in close proximity to other nerve cells, glands, or muscle cells capable of responding to the neurotransmitter.

Essentially all mammalian cells contain negatively charged proteins and ion transporter proteins in the plasma membrane, which cause the inside of the cells to have a net negative charge. Second, these transporter proteins maintain a differential in sodium and potassium concentrations such that the sodium ion concentration is greater outside than inside the cell. If conditions allow the sudden opening of membrane channels for sodium, a sudden influx of sodium briefly alters the polarity of the membrane in that location. This is called *graded potential* and specifically *excitatory potential* because depolarization moves the membrane closer to a threshold value that can open voltage regulated channels. If the impacted cell is another nerve cell, a frequent result of neurotransmitter secretion is to cause depolarization of this cell. Depending on the amount of neurotransmitter secreted or number of neighboring neurons also secreting neurotransmitter, this next cell may also exhibit an action potential if the stimulation is enough to open voltage-regulated channels. This is the essence of nerve impulses passing from neuron to neuron along a nerve fiber. However, not all neurotransmitters cause depolarization. Some neurotransmitters actually cause adjacent cells to become hyperpolarized. This serves to inhibit the activity of a given nerve tract. The type of action (depolarization or hyperpolarization) is a function of the neurotransmitter and the specificity of receptors on the surface of the neuron.

Autonomic Nervous System

Relative to lactation and the mammary gland, the nervous system can be simply divided into the central nervous system consisting of the brain and spinal cord and the peripheral nervous system consisting of the somatic nervous system under voluntary control and the autonomic nervous system under involuntary control. The autonomic nervous system contains two subdivisions. Sympathetic nerves (thoracolumbar division) exit the spinal cord primarily from the thoracic and lumbar vertebrae and the parasympathetic nerves from the head and neck or sacral region of the spinal cord (craniosacral division). Nerves from both divisions of the autonomic nervous system consist of two nerves per nerve tract. Anatomically, the majority of paired sympathetic nerves exit the spinal cord and synapse with their second neuron within clusters of paired ganglia located in a chain just outside the spinal cord (paravertebral ganglia). A second nerve exits the ganglion and travels to the structure ultimately innervated by the nerve. The first nerve of the pair is called a preganglionic nerve fiber and the second a postganglionic nerve fiber. The nerves of the parasympathetic division in contrast leave the spinal cord or cranial area and travel nearly to the final site of innervation before they synapse with their second neuron. The parasympathetic ganglia are typically embedded in the wall of the organ that is innervated. So anatomically, the preganglionic fibers of the sympathetic fibers are short, but the postganglionic fibers are relatively longer. The opposite is generally true for the parasympathetic division. In both divisions the nerve terminals of the preganglionic fibers secrete acetylcholine as the neuro-transmitter. The postganglionic fibers of the parasympathetic division also secrete acetylcholine, but those of the sympathetic division secrete norepinephrine. Nerve terminals that secrete acetylcholine are called *cholinergic fibers,* and those that secrete norepinephrine are called *adrenergic fibers.*

Milk Ejection and the Nervous System

Nerves supplying the mammary gland of the cow exit from the first to the fourth lumbar vertebrae and include sensory fibers and postganglionic sympathetic fibers. There is little credible evidence for innervation from parasympathetic fibers. In cows and other mammals, sensory receptors are abundant in the skin of the mammary gland and especially in the teat or nipple. In response to prepara-tion of the udder for milking or nuzzling of the offspring, nerve impulses travel via afferent nerves (branches of the inguinal nerve) to the dorsal root ganglia of the spinal cord and ultimately ascend the spinal cord along the dorsal funiculus to the midbrain. Branches project ultimately to impact clusters of nerve cells in the hypothalamus called *paraventricular* and *supraoptic nuclei.* These cells synthe-size oxytocin and a carrier protein neurophysin. The oxytocin and neurophysin are stored in the terminals of neurosecretory cells, which pass from the hypothal-amus to the posterior lobe of the pituitary gland. In response to neural stimulation of the mammary gland, the oxytocin is released into the bloodstream. It travels to the mammary gland, binds to receptors on the myoepithelial cells, and causes them to contract, thereby forcing milk from storage in the alveoli into the ducts to be harvested by the milking machine or suckling young. While the details were described only in the past 30 years, the essence of the milk ejection reflex has been recognized for thousands of years.

On the other hand, signals arising from the external environment, higher brain centers, or conditioned reflexes may also cause the release of oxytocin. It is com-mon to observe milk leaking from the teats of dairy cows as they enter the milk-ing parlor, long before the udder has been prepared for milking. Experimentally, direct stimulation of higher brain centers may also cause the release of oxytocin. These observations indicate that multiple neural pathways can act to modify the final neural pathway for release of oxytocin. Ultimately, associated nerve cells, which synapse with the neurosecretory cells of the paraventricular or supraoptic nuclei, act to either inhibit or facilitate this pathway. A predominance of cholinergic activity excites or facilitates, but stimulation of local adrenergic neurons impairs oxytocin secretion. It has long been recognized that stress at the time of milking interferes with the milk ejection reflex. Failure of oxytocin to be secreted as an explanation for impaired milk let down is called *central inhibition*. Since it is likely that stress causes stimulation of the sympathetic division of the auto-nomic nervous system, it is also possible that increased sympathetic nervous sys-tem–mediated vasoconstriction (α-adrenergic receptors) of the sphincters of the metarterioles in the mammary capillary beds might act to shunt oxytocin-laden blood away from the alveoli. Since the oxytocin acutely secreted at milking has a short half-life (~5 minutes), failure of delivery to the myoepithelial cells (so-called peripheral inhibition) may also explain failures of milk ejection. The often-repeated advice for careful, gentle handling of animals at milking or suckling is based on sound physiology.

The advent of radioimmunoassay (RIA) techniques and application to research with dairy animals was an important step in better understanding connections between physiological events, the neural system, and the endocrine system. Although many bioassays were critical in determining the broad, general concepts related to secretion of hormones, the ability to accurately measure rapid, acute changes in secretion of many hormones and growth factors in secretions from many animals was critical to understanding endocrine regulation of mammary function. Interestingly, measurement of oxytocin with RIA confirms bioassay data showing large variations in oxytocin response to milking or suckling. In goats and cows as many as 40 percent of the animals show no change in oxytocin with milk-ing stimulation. Moreover, milk yields of the mammary glands transplanted under the necks of goats were normal despite the lack of innervation or apparent milk ejection response. Perhaps for animals such as sheep and goats with large gland cisterns relative to mammary size, milk ejection is not essential for adequate milk removal. Figure 2.3 illustrates changes in blood oxytocin in response to 4 minutes of milking in two cows. One animal shows an abrupt increase in blood oxytocin, and the other cow essentially no response at all, yet milk yields were normal.

Clearly, there are species differences in the need or degree of oxytocin release at milking and details to be learned about other factors that may impact milk removal independent of oxytocin. Changes in sympathetic tone cause rhythmic contractions of the teats and tissue surrounding the gland cistern, and large ducts in the mammary glands of cows exhibit the presence of α_1-, α_2-, and β_2-adrener-gic receptors. Indeed ease of milking of cows has been associated with the den-sity of α-adrenergic receptors and the ratio between β_2 and α_2 receptors in mammary smooth muscle of the teat and in blood lymphocytes. It may be that receptor density on smooth muscle cells of the teat or in blood cells reflects recep-tor density for larger mammary ducts. Regardless, such findings suggest a degree of neurological input in control of the movement of milk from the alveoli to the gland cistern, which is independent of oxytocin (Lefcourt and Akers, 1983; Bruckmaier and Blum, 1998).

In summary, for the majority of species studied, initiation of the milk ejection reflex is likely essential to obtain milk. Sampling of milk from most nondairy

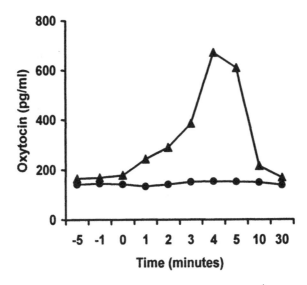

Fig. 2.3 Serum oxytocin measured in two cows in the period just before and after milking. For one cow (●) there is no measurable change in oxytocin concentration, but for the other cow (▲) oxytocin is increased four-fold within 4 minutes. Milk yields were normal for each animal. Adapted from Lefcourt and Akers, 1983.

species requires sedation and injection of exogenous oxytocin. Perhaps one of the consequences of genetic selection for milk production is the simultaneous selection for rapid milking and ease of milking. A reduction on dependence of the milk ejection reflex in these animals may be a consequence. In dual-purpose cattle a critical skill of the milker is to induce the cow to let down her milk. Primitive people used two primary means for this purpose. The first was to allow the calf to suckle a teat while the other teats were hand milked. In fact bringing the calf to the dam was often sufficient to induce milk let down. A second means to elicit milk ejection was to stimulate the vagina of the cow to be milked. The physiological basis for this success is now known to be induced secretion of oxytocin and minimization of stress to the lactating animal.

Endocrine System

No study of the mammary development or regulation of lactation can be complete without consideration of the endocrine system. The classic view of hormones was that these were substances produced by ductless glands for secretion into the bloodstream. Circulating hormones then traveled throughout the body to affect various target tissues. It is now clear that the distinction between hormones and various growth factors is muddled. For example, the liver is the major source for insulin-like growth factor I (IGF-I) in the circulation. Some portion of circulating IGF-I likely impacts mammary development, but IGF-I is also synthesized locally within the mammary gland. Systemically derived IGF-I fits the classic hormonal definition, but what about local tissue IGF-I? The molecule may have to diffuse only a few micrometers to find target cells. In fact, such local-acting molecules are described as autocrine acting when they impact the producing cells or paracrine if neighboring cells are stimulated. Moreover, neurotransmitters released into synaptic spaces between neighboring neurons or between nerve terminals and effector tissue mimic this form of local signaling. Whether the regulator is paracrine, synaptic, or endocrine in origin, only cells that express receptor proteins for the hormone or growth factor can respond. Regulators of the mammary gland are classified as mammogenic, lactogenic, or galactopoietic. Mammogenic hormones or growth factors have an effect on the growth and development of the mammary gland. Lactogenic molecules act to promote the structural and/or biochemical differentiation of the alveolar epithelial cells to synthesize and secrete milk. Galactopoietic agents function to maintain or enhance milk production once lactation is established.

Classification of Endocrine Glands and Secretions

Hormones and growth factors can be broadly classified based on their mechanisms of action. At the simplest level they either interact with target cell receptors located in the plasma membrane or with intracellular receptors. This is largely determined by the physical and chemical properties of the hormone. For example, the anterior pituitary hormone prolactin is a polypeptide of about 193 amino acids with a molecular weight of about 23,000. It is far too large to easily traverse the membrane of the mammary cell. Consequently, it like other protein hormones affects the mammary cells by first binding to receptors located on the plasma membrane. The binding reaction often initiates a series of biochemical events that ultimately change the intracellular concentration of potent regulators. Other hormones, or even the same hormone in a different target cell or under other conditions, might act by opening or closing an ion channel in the plasma membrane. The change in internal ion concentration is thus responsible for the effect attributed to the hormone.

Homeostasis is possible only because of the functional coordination between the nervous and endocrine systems. For example, in support of lactation, neural signals induce the secretion of oxytocin from the posterior pituitary to allow the suckling young to obtain milk stored in the mammary gland. As it is with various physiological functions (e.g., regulation of feeding behavior, blood pressure, respiration), the hypothalamus controls secretion of anterior pituitary hormones. Either directly or indirectly a very large number of hormones regulates mammary development and function (Table 2.1).

Given the overriding importance of the anterior pituitary hormones and either their direct (e.g., prolactin) or indirect effects (e.g., follicle-stimulating hormone [FSH] and estrogen production) on the mammary gland, it is worth considering the relationship between the hypothalamus, brain, and pituitary gland. The ante-

Table 2.1 Major hormones affecting mammary gland development or function

Endocrine gland	Hormone secreted	Major mammary effect
Anterior pituitary	Adrenocorticotropic hormone (ACTH)	Stimulates adrenal gland secretion of cortisol
	Follicle-stimulating hormone (FSH)	Estrogen secretion
	Growth hormone (GH)	Stimulates milk production
	Luteinizing hormone (LH)	Progesterone secretion
	Prolactin (Prl)	Lactogenesis, cell differentiation, milk protein gene expression
	Thyroid-stimulating hormone (TSH)	Stimulates thyroid gland to secrete thyroxine and triiodothyronine
Posterior pituitary	Oxytocin	Milk ejection reflex
Hypothalamus	Growth hormone–releasing hormone	Stimulates GH secretion
	Somatostatin	Inhibits GH secretion
	Thyrotropin-releasing hormone	Stimulates TSH secretion (as well as Prl and GH)
	Corticotropin-releasing hormone	Stimulates ACTH secretion
	Prolactin-inhibiting hormone (dopamine)	Inhibits Prl secretion
Thyroid	Thyroxine; triiodothyronine	Stimulates oxygen consumption, protein synthesis
	Thyrocalcitonin	Calcium and phosphorus metabolism
Parathyroid	Parathyroid hormone	Calcium and phosphorus metabolism
Pancreas	Insulin	Glucose metabolism (species variation)
Adrenal cortex	Glucocorticoids (cortisol, corticosterone)	Lactogenesis, cell differentiation, milk protein gene expression
Adrenal medulla	Epinephrine	Inhibition of milk ejection reflex (peripheral)
Ovary	Estrogen	Mammary duct growth
	Progesterone	Mammary lobulo-alveolar development, inhibition of lactogenesis
Placenta	Estrogen Progesterone (species dependent)	See Ovary
Placental lactogen		Mammary development

rior pituitary gland is not directly innervated; instead, hormones secreted by neurons of the hypothalamus enter the hypophyseal portal blood supply, which drains directly to the anterior pituitary gland. This unusual anatomical arrangement of blood vessels allows small quantities of hormones produced by the hypothalamic neurons to regulate the activity of the anterior pituitary cells. This is important because it is likely that if these hypothalamic regulators were delivered to the general circulation before reaching the pituitary, the dilution effect would render them ineffective. The hypothalamus secretes stimulatory or inhibitory hormones for each of the major anterior pituitary hormones. The relationship between the hypothalamus and the pituitary gland is illustrated in Figure 2.4. However, it should not be forgotten that the activity of the clusters of nerve cell bodies (nuclei) of the hypothalamus is also influenced by neurotransmitters from other higher-brain areas. Finally, secretion of the pituitary hormones is generally regulated by a series of negative feedback loops. For example, increased concentrations of thyroid-stimulating hormone (TSH) may act directly on the basophiles of the anterior pituitary to suppress further secretion of TSH. Such a path is sometimes called a short, short negative feedback loop. Increased secretion of triiodothyronine in response to TSH might negatively feed back at the level of the pituitary or hypothalamus for homeostatic control of circulating TSH. Similar scenarios can be deduced for many of the other pituitary hormones.

Mechanisms of Hormone Action

Cell surface–acting hormones interact with one of three known classes of receptor proteins defined by the signal transduction method that is employed. Ion channel–linked receptors are involved in rapid signaling usually between electrically excitable cells. The various neurotransmitters account for most of these examples. A second type, the ubiquitous G protein–linked receptors, function by modifying the activity of a separate plasma membrane–anchored protein, which may be an enzyme or may act as an ion channel. The interaction between the receptor (usually only after having been bound with hormone) and the anchored membrane protein is in turn mediated by the action of a third protein, a trimeric GTP-binding protein or G protein. If this target protein is the enzyme type, activation alters the concentration of one of a number of intracellular mediators (e.g., cyclic GMP, cyclic AMP, diacylglycerol, inositol triphosphate). It is important to understand that G protein hormone-mediated effects on target cells can be inhibitory or stimulatory. For example, the illustration in Figure 2.5 depicts the response if a stimulatory G protein variant (G_s) is activated in response to hormone binding. The result is an increase in the intracellular concentration of a second messenger molecule.

Let's suppose, however, that the illustrated target cell also has receptors for another hormone whose effect is mediated by an inhibitory G protein variant (G_i) and that the second messenger molecule is the same. In this instance hormone binding to the receptor would activate G_i with the effect of inhibiting the production of the second messenger. This suggests that the overall response of the cell to these two hormones would depend on the relative concentration of the two receptors on the cell surface and ratio of circulating concentrations of the hormones in the blood nourishing the target cells. Overlap in signaling pathways, changes in secretion of hormones, and alterations in expression of cell receptors provide

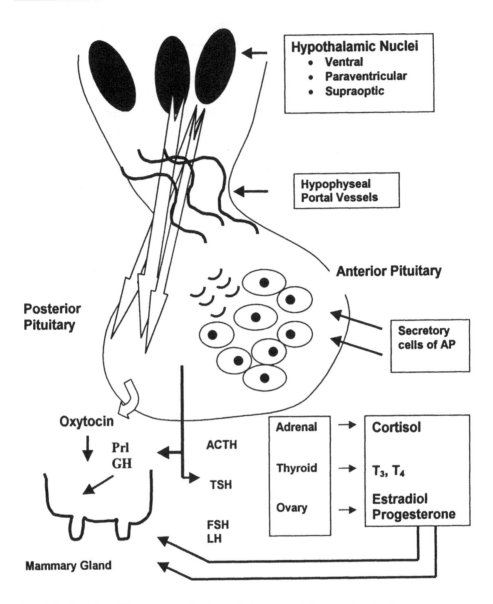

Fig. 2.4 Relationships between the hypothalamus and the anterior pituitary and posterior pituitary glands and control of hormone secretion are illustrated. Large arrows indicate neurosecretory cells passing from the supraoptic and paraventricular nuclei to the posterior pituitary for secretion of oxytocin. Neurosecretory cells which secrete releasing or inhibiting hormones into the hypophyseal portal vessels, impact the anterior pituitary. Anterior pituitary hormones directly (Prl and GH) or indirectly (ACTH, TSH, FSH, and LH) produce mammary active hormones (cortisol, T_3, T_4, estradiol, progesterone). Adapted from Tucker, 1985.

many opportunities for regulation of cell response to hormones or growth factors. Common membrane-anchored enzymes regulated by G proteins include adenylate cyclase and phospholipase C. Changes in activity state of adenylate cyclase

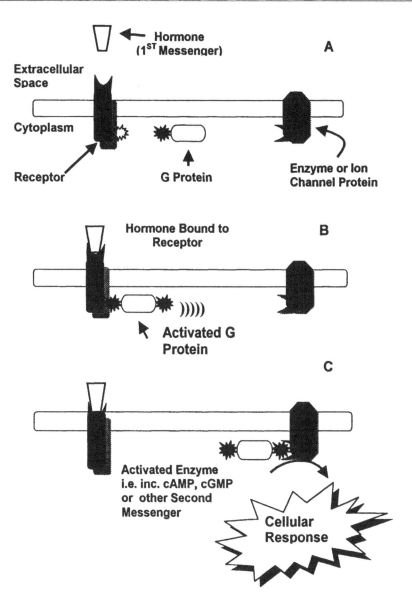

Fig 2.5 Example of a G protein–linked hormone mechanism of action is shown. Activation of the receptor is induced by the binding of hormone (1st messenger) (**A**). This leads to a conformation change in the cytoplasmic domain of the receptor to allow binding of a complementary G protein and activation (**B**). This activated complex links to a second membrane-bound protein (enzyme or ion channel), which is responsible for the effect associated with hormone actions (**C**).

as impacted by hormone activation of G_s or G_i proteins influence the capacity of the enzyme to convert ATP to cyclic AMP (cAMP). Cyclic AMP then binds to a cytosolic protein, cyclic AMP-dependent protein kinase (A-Kinase). This binding allows regulatory proteins (subunits) to detach from inactive A-Kinase so that the formerly inactive catalytic subunits of the complex are then activated. These active enzymes then elicit the phosphorylation of proteins unique to particular target cells. These phosphorylated proteins have potent biochemical effects including activation of other enzymes or gene activation. For example, the regulatory noncoding sequences of some genes contain a cyclic AMP response element (CRE). The binding of a specific regulatory protein called a CRE-binding (CREB) protein, when the protein is phosphorylated, activates this region of the gene. This is tied together by the fact that cAMP-activated A-Kinase phosphorylates CREB.

Other G proteins function by activating (or inhibiting) the inositol phospholipid-signaling pathway. As usual the effects are initiated by hormone binding to its receptor, which causes it to bind to an inactive G protein. Binding of GTP displaces GDP, and the G protein is activated. Activated G protein translocates within the lipid bilayer of the plasma membrane and binds to and activates phospholipase C. The enzyme hydrolyzes the phospholipid phosphatidylinositol bisphosphate (PIP_2) to produce diacylglycerol and inositol triphosphate (IP_3), both of which may act as second-messenger molecules. Diacylgylcerol activates specific protein kinases (analogous to the effects of cAMP), and IP_3 triggers the release of Ca^{2+} ions. Increased Ca^{2+} may stimulate activity of specific enzymes directly to act as a second messenger or can bind to the regulatory protein calmodulin to impact cellular function. To finish the discussion of G proteins, if the target protein acts as an ion channel, activation acts to change permeability of the membrane for that ion. The G protein–linked receptors make a very large family of surface receptors including the oxytocin receptor.

A third type, the enzyme-linked receptors, act directly as enzymes or are linked to enzymes following hormone binding. Many of these hormone receptors are protein kinases themselves. These include the serine/threonine and tyrosine kinases. Essentially, the binding of the hormone to the receptor activates the receptor (or directly linked protein) so that it is capable of phosphorylating cellular proteins. Phosphorylation is associated with altered activity of the affected protein. The mammary active regulators IGF-I, insulin, EGF, and vascular endothelial growth factor (VEGF) are all examples of tyrosine-specific protein kinases (Butler et al., 1998; Adams et al., 2000).

Interestingly, mechanisms of action for prolactin and growth hormone, two protein hormones with long recognized importance in mammary development and function, have only recently been elucidated. Additional complexity exists because of the presence of multiple forms of the receptors of each hormone. Both receptors are members of the structurally related cytokine receptor superfamily, which includes several interleukins, erythyropoietin, and leptin among others. Both prolactin receptor (PrlR) and growth hormone receptor (GHR) are single-chain proteins with a single transmembrane domain so that the protein spans the entire plasma membrane similar to many of the growth factor receptors. However, the receptors are not active as kinases themselves. Hormones in the cytokine family have two sites capable of binding to their receptor proteins. Initially a hormone binds (site 1) to a receptor to create an inactive complex. This hormone-receptor complex diffuses within the membrane to bind with another receptor protein via the second binding site, which leads to receptor homodimerization and formation

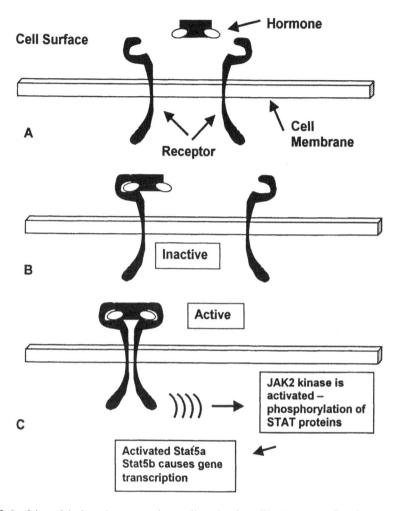

Fig. 2.6 Ligand-induced receptor homodimerization. The hormone first interacts with the receptor to form an inactive complex (**A**). The hormone then binds to a second receptor (**B**) to form a dimer and an active complex (**C**). Formation of the complex allows the activation of JAK2 kinase, which then phosphorylates Stat proteins. Stat5a and Stat5b are closely involved in Prl stimulation of specific milk protein gene transcription. Receptor binding can also stimulate additional signaling pathways, including mitogen-activated protein (MAP) kinase and protein kinase C (PKC) (Das and Vonderhaar, 1997; Henninghausen et al., 1997).

of an active complex (Fig. 2.6). Studies with mutant forms of the Prl or growth hormone (GH) confirmed the importance of homodimerization in hormone action. Specifically, versions of the Prl or GH with impaired binding sites are unable to form homodimer pairs with their respective receptor proteins and are devoid of activity. Although it had been known for some time that stimulation of target cells with Prl or GH caused tyrosine phosphorylation of a number of cellular proteins, the cytoplasmic domains of the receptors have no enzymatic activity.

This means that hormone binding and dimer formation have to activate other cellular kinases. A breakthrough to understanding this signaling cascade came with the discovery that Janus tyrosine kinase 2 (JAK2) was activated with hormone stimulation. The kinase belongs to a family with four members: JAK1, JAK2, JAK3, and Tyk2, all of which appear to be involved in signaling of cytokine receptors. For PrlR and GHR, JAK2 seems especially important. For PrlR, JAK2 is constitutively associated with the receptor, but with GHR the enzyme associates with the receptor only after hormone binding and dimer formation (Goffin and Kelly, 1997; Hynes et al., 1997).

The signaling pathway is thought to involve JAK2-induced phosphorylation of a transcription factor known as signal transducer and activator of transcription (STAT). One of these, STAT5, is specifically implicated in Prl stimulation of the casein gene. Yang et al. (2000a) have recently shown in both rat and bovine mammary explant cultures that additions of GH, Prl, or IGF-I stimulated STAT5 DNA binding activity. In particular, STAT5 activity was detectable in extracts prepared from snap frozen biopsy material. However, when explants were incubated for 1 hour in the absence of lactogenic hormones, STAT5 activity was not detectable but inclusion of either Prl or GH stimulated the appearance of the STAT5-DNA binding complex in an additive fashion. Addition of IGF-I also stimulated STAT5 binding activity. On the other hand hormone additions in the short-term did not alter expression of the STAT5 protein. This suggests that acute regulation involves protein activation or binding rather than protein synthesis. This was further shown in experiments in which cycloheximide addition was used to block protein synthesis during the period of treatment with Prl or GH; specifically inclusion of cycloheximide had no effect on ability of the tissue to response to either hormone. In a subsequent study Yang et al. (2000b) showed that STAT5 activity was readily detected in mammary tissue of lactating cows but not in the mammary tissue of nonlactating cows. Furthermore, STAT5 activity increased with milking frequency and was moderately correlated ($r = 0.51$) with milk protein percent. Somewhat unexpectedly, long-term treatment of lactating cows with GH or GH-releasing factor (GRF) raised milk production but depressed STAT5 activity in the mammary gland. In contrast, these treatments had no effect on abundance of STAT5 protein in the mammary gland. These results support the idea that STAT5 activation and protein synthesis are affected by physiological status and that hormones known to impact differentiated mammary function (Prl, GH, and IGF-I) may all impact the mammary epithelium by a common signaling pathway involving this regulatory protein.

Overall, those hormones that act by binding to cell surface receptors are described as having second messenger mechanisms of action. The explanation for this terminology is that the hormone itself is considered the "first" messenger and the critical molecule(s) influenced inside the cell is the "second" messenger(s). Although steroid hormones, thyroid hormones, retinoids, and vitamin D are structurally diverse, these relatively small hydrophobic molecules all pass the plasma membrane of target cells to act internally. These structurally similar receptors make up the steroid hormone superfamily. The hormone binds to its receptor in the nucleus. Binding releases associated proteins called *chaperones* so the newly formed hormone-receptor complex can attach to a specific region of the DNA. This association allows transcription of the gene(s) adjacent to this binding site. Thereafter, transcription of the specific gene(s) occurs, and the new mRNA is processed and transported to the ribosome for translation. These newly minted proteins are responsible for the hormone effects observed (Fig. 2.7). A general rule is that responses to these hormones are slower than for the surface-acting peptides. This seems logical given the differences in mechanisms of action. Estrogen, progesterone, cortisol, triiodothyronine, and retinoids are all potent stimulators of mammary cells (Hansen and Bissell, 2000; Rosen et al., 1999).

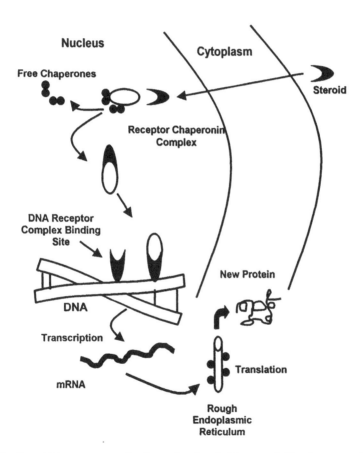

Fig. 2.7 Steroid hormone mechanism of action is illustrated. The hormone diffuses into the nucleus and binds to a receptor-chaperonin complex. Binding allows the chaperone proteins to dissociate, and the activated hormone receptor combination binds to specific sites on the DNA. This triggers the transcription of certain genes and production of new mRNA. The new mRNA is translated and new proteins made. These new proteins are responsible for the biological effects of the hormone.

Chapter 3
Functional Development
of the Mammary Gland

Onset of Milk Secretion

The capacity for milk production to occur at the onset of parturition requires a great deal of physiological coordination. Appropriate development of mammary alveoli, timely structural and biochemical differentiation of the alveolar secretory cells, metabolic adjustments to supply needed substrates, and regular milking or sucking must occur. Successful lactation requires at least three distinct events: (1) the prepartum proliferation of alveolar epithelial cells, (2) biochemical and structural differentiation of these cells, and (3) synthesis and secretion of milk constituents.

Lactogenesis

Lactogenesis is usually described as a two-stage process. Stage 1 consists of limited structural and functional differentiation of the secretory epithelium during the last third of pregnancy. Stage 2 involves completion of cellular differentiation during the immediate periparturient period coinciding with onset of copious milk synthesis and secretion. During lactation, the secretory cells synthesize and secrete copious amounts of carbohydrate, protein, and lipid. Production of this complex mixture of nutrients depends on coordination between biochemical pathways to supply metabolic intermediates and secretory pathways for secretion. For example, the disaccharide lactose is the predominant milk sugar. The enzyme complex necessary for lactose synthesis, membrane-bound galactosyltransferase and the whey protein α-lactalbumin, combine in the Golgi apparatus to form lactose synthetase, which links glucose and galactose to produce lactose. Activation of the α-lactalbumin gene and synthesis of α-lactalbumin is most closely associated with stage 2 of lactogenesis.

Cellular Differentiation

Certainly, biochemical differentiation of the secretory cells is required for onset of milk secretion. However, the cells must also acquire the structural machinery needed to synthesize, package, and secrete milk constituents. When alveolar cells first appear during midgestation, they exhibit few of the organelles needed for copious milk biosynthesis or secretion. The cells are characterized by a sparse cytoplasm with few polyribosomes, a few clusters of free ribosomes, limited rough endoplasmic reticulum, rudimentary Golgi usually in close apposition to the nucleus, some isolated mitochondria, and widely dispersed vesicles. Individual cells often contain large lipid droplets (especially during later stages of gestation) that, along with irregularly shaped nuclei, account for much of the cellular area (Fig. 3.1). Soon after the alveoli appear, lumenal spaces accumulate increasing volumes of fluid containing serum-derived proteins. These secretions result in formation of immunoglobulin-rich colostrum, which—depending on the species—may be essential for the survival of the offspring. Electron microscopy studies showed the dramatic structural changes in the alveolar secretory cells at the onset of lactation (Fig. 3.2).

Classic mammary explant culture studies demonstrated that the major positive regulators of structural differentiation of the secretory cells are glucocorticoids and prolactin. (Details of hormone changes and endocrine regulation of mammary development and lactogenesis appear in a subsequent section.) Mammary tissue explants from pregnant or steroid hormone–primed donors exhibit both biochemical and structural differentiation when incubated in culture medium containing a combination of insulin, glucocorticoids, and prolactin. While it is widely believed that insulin is necessary for mammary tissue maintenance in culture, species variation in insulin sensitivity of the mammary gland in intact animals casts some doubt on this belief. Recent data support the idea that insulin-mediated effects on mammary cells in culture may actually represent effects more appropriately ascribed to the insulin-like growth factors (IGF-I and IGF-II). This is because mammary epithelial cells have specific IGF-I receptors, and insulin (especially at higher concentrations typical of culture experiments) is likely to bind to the IGF-I receptor. In general terms, glucocorticoids are most closely associated with development of rough endoplasmic reticulum and prolactin with maturation of the Golgi apparatus and appearance of secretory vesicles. Prolactin added to explant cultures incubated with insulin and glucocorticoids dramatically increase the de novo synthesis and secretion of α-lactalbumin and caseins (Rosen et al., 1999; Goodman et al., 1983).

Cell-Cell Interactions

During the second stage of lactogenesis, the approach of parturition signals both the structural differentiation of the alveolar cells and maturation of tight junctions (zona occludens) between the cells. Increases in circulating glucocorticoids along with declining progesterone seem to be especially important. Once this occurs,

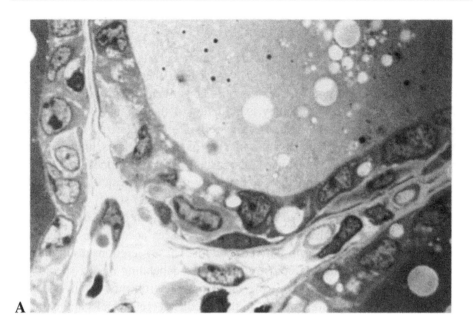

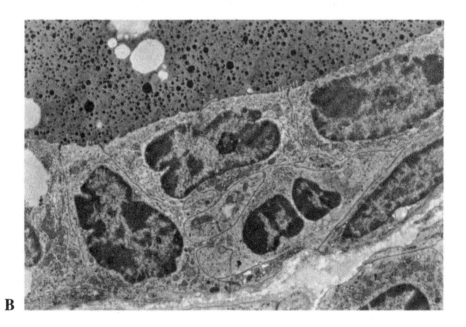

Fig. 3.1   Shown are companion light (**A**) and transmission electron (**B**) microscopic images of bovine mammary tissue of a nonlactating cow in late gestation. Note the relatively large proportion of cell area occupied by the nuclei of the cells, relative lack of cellular organelles, absence of cellular polarity and minimal evidence of secretion (unpublished micrographs).

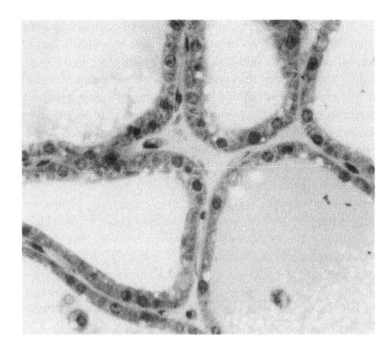

A

B

Fig. 3.2   Illustrated is an example of a light microscopic (**A**) and a transmission electron microscopic image (**B**) of lactating bovine mammary tissue. Note the alveoli with rounded, basally displaced nuclei, scattered fat droplets, and evidence of secretions (**A**). Note the lacy appearance of the apical ends of the well-differentiated cells of the alveoli, darkly stained basolateral areas, and highly polarized cells of the alveoli. Confirmation that the lacy appearance indicates the presence of abundant secretory vesicles is evident in the electron microscopic view (**B**). **A** is an unpublished image; **B** is from Nickerson and Akers, 1984.

generally just after parturition, paracellular transport of components between the cells is dramatically reduced. This creates an effective blood-milk barrier so that transfer of serum components into milk or milk constituents into blood is minimized. This does not mean that transport cannot occur but that wholesale leakage is prevented. The effectiveness of this barrier function is readily apparent from the study of secretions obtained from animals with acute mastitis or experimental treatments known to disrupt the tight junctions. One of the effects is the appearance of serum proteins in milk (e.g., albumin). Conversely, these situations also allow abrupt increases in the appearance of lactose and α-lactalbumin (and likely other milk components) in serum.

The initial stages of secretory cell differentiation can actually be monitored indirectly by taking advantage of these immature "leaky" tight junctions (McFadden et al., 1987). Specifically, as small quantities of milk proteins or other constituents are synthesized and secreted into the lumenal spaces of the alveoli, some of these specific milk components appear in blood. For example, in pregnant heifers, α-lactalbumin concentrations in serum rise from almost undetectable levels prior to 160 days prepartum to reach a plateau between day 120 and 50 prepartum. Concentrations then increase very rapidly during the last 3 weeks prior to calving (Fig. 3.3).

This pattern strongly supports the two-stage theory of lactogenesis that was initially proposed based on changes in the ability of isolated, incubated mammary tissue to synthesize milk components (Fig. 3.4). Similar changes in serum concentrations of β-lactoglobulin and lactose have also been noted. Furthermore, during the immediate periparturient period, serum α-lactalbumin concentrations peak just after calving but very rapidly decline. This abrupt decrease is believed to indicate the rapid maturation of the tight junctions at this time (Fig. 3.5). Changes in serum α-lactalbumin can also be used to monitor mammary development during hormonal induction of lactation.

During established lactation, measurement of milk components in serum can be used to indirectly monitor mammary function. For example, concentrations are likely impacted by changes in intramammary pressure and udder health. Early studies suggested an association between presence of lactose in blood and milking interval. Over the course of lactation, serum α-lactalbumin concentrations are greater in cows milked 2 times compared with those milked 3 times per day (102 vs. 73 ng/ml). Changes in udder health as indicated by milk somatic cell count (MSCC) correspond with changes in serum α-lactalbumin. Concentration averaged 129 ng/ml for cows with low MSCC (<25,000 cells per ml) but was markedly higher (347 ng/ml) for cows with high MSCC (>500,000 cells per ml). Averaged across stage of lactation, lactation number, and between two herds, serum α-lactalbumin was positively correlated with MSCC. Intramammary infusion of *Escherichia coli* endotoxin also induces a rapid but reversible increase in MSCC and corresponding short-lived increases in serum α-lactalbumin. These results support the idea that high MSCC causes increased diapedesis of leucocytes into milk, disruption of tight junctions, and consequently greater passage of milk components into blood.

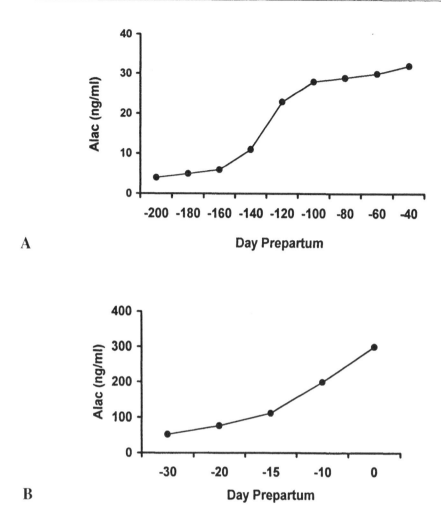

Fig. 3.3 Changes in serum concentrations of α-lactalbumin (Alac) are shown for Holstein heifers between 210 and 40 days before calving (**A**) and between 30 days before and day of calving (**B**). Little α-lactalbumin is detected until about 140 days before calving when concentrations increase to about 30 ng/ml. Concentrations remain near this value until the month before calving when there is a further dramatic increase. These two periods of increase mirror the two stages of lactogenesis. Adapted from McFadden et al., 1987.

Role of Milk Removal in Milk Secretion

As discussed briefly with respect to mammary involution, once lactation is established regular milking or suckling is essential for maintenance of lactation. Although increased alveolar pressure likely affects blood supply and contributes to impaired milk secretion in the absence of milking, several intriguing studies suggest that local regulation of milk secretion by milk removal is also chemically

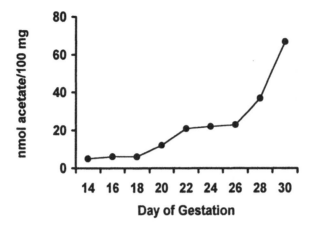

Fig. 3.4 Incorporation of acetate into mammary slices from pregnant rabbits is shown. Incorporation is initially very low but is increased after day 19 of gestation. Rates remain relatively constant until just before parturition on day 30. Incorporation of glucose into lactose follows a similar pattern, except the abrupt second increase occurs closer to parturition (curve shifts to the right). Like changes in serum α-lactalbumin in cows, this pattern supports a two-stage model for lactogenesis. Data adapted from Strong and Dils, 1972.

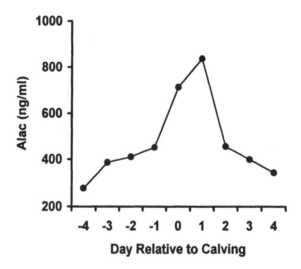

Fig. 3.5 Data illustrate changes in serum α-lactalbumin (Alac) around the time of calving in primiparous Holstein heifers. The marked increase close to calving is believed to represent the rapid onset of lactogenesis and the decrease after the maturation of tight junctions between alveolar epithelial cells as normal milking begins. Data adapted from McFadden et al., 1987.

mediated. Specifically, studies in goats led to the isolation and partial identification of a protein fraction called FIL (feedback inhibitor of lactation). Preparations of FIL were shown to inhibit milk secretion both in vivo and in vitro in a concentration-dependent manner. In vitro studies of milk protein secretion suggest that there are two regulatory pathways: a so-called regulated pathway stimulated by an increase in intracellular calcium concentrations and also a constitutive secretion independent of calcium. FIL apparently acts primarily by blocking constitutive secretion of the alveolar epithelial cells. Based on studies of milk protein secretion by isolated mammary acini, none of the tested known signal transduction pathways appear to explain FIL's mechanism of action (i.e., activation of G-proteins, or kinase C). In addition to factors acutely affecting milk secretion, increased milking frequency is associated with an increased gland size and apparent mass of secretory tissue in unilaterally milked goats. It may be that removal of FIL or other mediators can also induce mammary cell growth during lactation. However, it is also possible that differences in mass reflect a lesser rate of secretory cell loss with frequent milking (Peaker and Wilde, 1996).

The stimulatory effect of milking or secretion removal on the secretory cells is apparent among animals prepartum milked. For example, when udder halves of cows were treated so that one-half of the mammary gland was hand milked beginning 2 weeks prior to calving, development of the prepartum-milked glands was markedly accelerated. Composition of secretions from the prepartum-milked glands was initially colostrum-like, but for most animals within several days of milking (prior to parturition), the composition of secretions closely mirrored that of normal milk. When tissues were sampled shortly after calving, secretory cells from prepartum-milked glands contained a larger proportion of well-differentiated cells than for alveoli from nonprepartum–milked glands. This simple experiment suggests that removal of secretions can have a dramatic impact on lactogenesis irrespective of the overriding influence of periparturient secretion of lactogenic hormones (Akers et al., 1977; Akers and Heald, 1978).

Milk Component Biosynthesis

The mammary gland is an unusual exocrine gland in several respects. The product is a very complex mixture, which depends on apocrine and meocrine modes of secretion. Other components are derived by passage of soluble molecules across (transcellular) and sometimes between (paracellular) the cells. Physically milk is a complex solution of salts, carbohydrates, miscellaneous compounds with dispersed proteins and protein aggregates, casein micelles, and fat globules. Milk osmolarity generally equals blood (~300 mOsm) and has a pH between 6.2 and 7.0. Bovine and human milk average pH 6.6 and 7.0, respectively.

Once initiated, milk secretion continues more or less continuously throughout lactation. Milk is stored within the lumen of the alveoli and ductular system until it is removed by the milking machine or the suckling offspring. Interestingly, suckling intervals vary widely between mammals, ranging from minutes to hours in cattle, to once daily in rabbits, to once every 2 days in tree shrews, or only once a week in some seals. Moreover although there are species-specific changes in milk composition with stage of lactation, milk composition for most of lactation is generally only moderately affected by environmental or nutritional changes, despite the often dramatic changes in milk volume. Function of the mammary gland during established lactation is closely linked with a number of hormones, growth factors, and local tissue regulators (Table 2.1), but it is difficult to ascribe a specific transport activity to a particular molecule or to determine if effects are direct or indirect. In large part this is because no available in vitro system is an adequate model for synthesis, secretion, and storage of true milk with its complex composition. Study of systems that allow isolated cells, organoids, or tissue fragments to synthesize and secrete particular milk constituents (e.g., caseins, α-lactalbumin, or lactose) may be adequate to elucidate selected biochemical regulatory pathways but certainly not regulation of lactation.

Any discussion of milk synthesis also has to consider not only the nutrients necessary for milk synthesis but also the physiological adjustments necessary to supply these nutrients to the mammary gland. Requirements for high levels of milk production are staggering at first glance. For example, in the dairy cow the energy requirements for milk production can easily approach 80 percent of net energy of intake. Demands for lactose production can easily require 85 percent of circulating glucose. It has been calculated that high-producing cows must mobilize adipose tissue and body nutrient reserves equal to approximately one-third of the milk produced during the first month or more of lactation. Clearly, finely tuned coordinated interactions between all the major physiological systems are necessary for success. Second, metabolic pathways for the synthesis and secretion of milk protein, fat, and carbohydrate must be explained. Lastly, an appreciation of the cellular mechanics of synthesis and secretion completes the story.

Structure and Function at the Secretory Cell Level

A common biological theme is that structure follows function. The structural differentiation of the alveolar cells around the time of parturition illustrates this principle very well. As described in previous sections (e.g., "Cytology Indices" in Chapter 1), these cells become polarized, with appearance of abundant arrays of rough endoplasmic reticulum (RER), numerous mitochondria, and competent tight junctions just as full-scale milk synthesis and secretion is initiated. Figure 3.2 illustrates the major structural elements of an actively secreting mammary cell. The fine structure of the mammary epithelium during lactation has now been documented in many species. This generally uniform structure includes basal and paranuclear cytoplasms occupied by parallel arrays of RER. The supranuclear Golgi apparatus typically consists of stacks of smooth membranes whose terminal cisternae release casein and lactose containing secretory vesicles. Lipid droplets and secretory vesicles seemingly fill the apical ends of the cells, and microtubules are most frequently observed oriented perpendicular with respect to the apical plasma membrane. Mitochondria and free ribosomes are abundant throughout the basal-lateral cytoplasm. The basal plasma membrane is often thrown into complex folds, believed to indicate active pinocytosis. Myoepithelial cells frequently occur interspersed between the basal plasma membrane and the basal lamina that forms a loose barrier between the simple cuboidal epithelial cells and the underlying stromal tissues. The dramatic alterations in epithelial cell ultrastructure between prepartum and postpartum periods are illustrated in Figures 3.1 and 3.2.

In the intact mammary gland, five routes of secretion across the mammary epithelium have been described: these are (1) membrane route, (2) Golgi route, (3) milk fat route, (4) transcytosis, and (5) the paracellular route. The membrane route refers to interstitial fluid–derived substances that cross the basolateral membrane, traverse the cell, and pass across the apical membrane into milk. Examples are water, urea, glucose, and some ions. To utilize the Golgi route, these products are synthesized, sequestered, or packaged into secretory vesicles that bud from the stacks of Golgi membranes. These vesicles either individually or in chains fuse with the apical plasma membrane to release their contents to become part of milk. Interestingly, this mode of secretion involves the cytoskeleton and especially the microtubules since drugs that block tubulin formation (the major protein component of microtubules) reversibly block secretion. Examples of these products are lactose, caseins, whey proteins, citrate, and calcium. The milk fat route refers to substances that become entrained with the budding lipid droplets as they are released from the apical cell surface to become part of milk. Actually, as the fat droplets are secreted, it is not uncommon for bits of cytoplasm to become engulfed by plasma membrane and secreted from the cell. Furthermore, the number of these secreted cellular fragments appears to vary substantially between species. For example, they are especially common in the milk of goats. The apparently random nature of the cytoplasmic portion that is lost in this manner means that virtually any substance or cytoplasmic organelle could be secreted this way. The variety in fragments secreted in this manner is illustrated in Figure 3.6, an example taken from the alveolar lumen of the mammary tissue of a lactating cow.

Other lipid soluble molecules also can be secreted because of their affinity for lipid or as part of milk fat globule membrane (Mather, 2000). Examples include a variety of hormones and drugs. In transcytosis, vesicles derived from the basolateral membrane (pinocytosis or endocytosis) are transported in membrane-bound vesicles for exocytosis at the apical membrane. Finally, in the paracellular route there is direct passage for materials in the interstitial fluids between the epithelial cells and into milk. Except in situations of disease (i.e., mastitis or failure of frequent milk removal), the paracellular route is likely of minimal importance during established lactation. However, it should be remembered that even during normal lactation in the absence of known disease some milk constituents (i.e., lactose and whey proteins) can be detected in blood serum. In the absence of misdirected secretion of secretory vesicles (for which there is little evidence), apparent movement of specific milk components from storage in the alveolar lumen suggests paracellular transport

Fig. 3.6 Transmission electron microscopy image of a photomicrograph of milk in the alveolar lumen of a lactating cow is shown. To the lower left is a fat droplet and associated cytoplasmic fragment. Within the fragment there are several fragments of rough endoplasmic reticulum and several secretory vesicles. In the surrounding space there are numerous casein micelles (dense black structures). Taken from Akers, 1991.

is not zero even in established lactation. Some of this might reflect transport associated with temporary breaks in epithelial integrity during diapedesis of leucocytes. Indeed, there is a positive correlation between somatic cell count (essentially a measure of leucocytes) in milk and the concentration of α-lactalbumin in blood serum. Regardless, permeability of the mammary epithelium by this pathway is markedly reduced around the time of parturition. There is also evidence that maturation of the tight junctions is stimulated by glucocorticoids as shown by increased transepithelial electrical resistance and reduced transfer of inulin across confluent monolayers of murine mammary cells. These effects fit with well-characterized changes in periparturient hormone secretion. Specifically, it is suggested that progesterone and possibly locally produced TGF-β and prostaglandin F2α (PFG$_{2\alpha}$) act to inhibit tight junction formation during pregnancy but that the decrease in prepartum progesterone and corresponding increase in cortisol promote the final maturation of these junctions. However, even during established lactation, activity of the paracellular secretion route can affect milk composition. Disruption is evident during mastitis as well as when supraphysiological doses of oxytocin are given to induce milk letdown. Since most nondairy animals must be injected with oxytocin to obtain milk samples, it is questionable if the composition of these samples is truly accurate. This is likely especially true for the aqueous phase of milk.

Milk is a complex fluid with many hundred individual soluble minor components; however, the major nutritive value of milk is accounted for by its gross composition of fats, proteins, carbohydrates, ions, and water. Water is typically determined as loss in weight by drying under conditions to minimize loss of organic materials. Fat is most often determined by extraction with organic solvents and is largely the triacylglycerols of the milk fat droplets. Protein content of milk is often estimated by analyzing the milk for nitrogen content and multiplying the result by a factor, usually 6.38, which estimates the average nitrogen content of proteins. However, colorimetric procedures are also often used in research laboratories. These methods depend on the capacity of the proteins to bind cationic dyes. Protein contents of unknown samples are estimated by comparison with the dye binding for a standard curve of purified protein. Milk proteins are also often divided into two broad classes or fractions, the caseins and the whey proteins. Whey proteins remain soluble, but the caseins precipitate at a pH of about 4.6. Milk carbohydrate is usually expressed as lactose, but older analytical methods may mistake lactose for other complex carbohydrates. This is of particular concern for those species that produce little if any lactose. Because gross milk composition is economically important in the dairy industry, routine milk testing of dairy cows now most often depends on rapid, automated techniques that estimate milk fat, protein, and lactose in single samples based on the ability of these components to absorb infrared light at 5.8, 6.5, and 9.6 nm, respectively.

The combination of minerals in milk, often expressed as the ash content, is calculated from the weight of residue remaining after incineration. This is really an inexact gross measure of mineral content since most organic ions are destroyed by high heat. Moreover, much of the phosphate in the ash is derived from the phosphorus of proteins and phospholipids. Carbonate, chloride, sodium, and potassium may also be volatilized by ashing. Despite limitations of sampling methods between species, and use of various assay methods, when combined with estimates of milk yield, measures of gross milk composition provide valuable information on the nutritive value of milk to the developing offspring. A cautionary note is advised; milk composition data are available for only about 225 of the known 4,000 species of mammals. Secondly, the data for these various mammals range from voluminous for dairy species to only single samples for some animals. Regardless, interspecies variation is very large: fat percentage ranges from a trace (zebra) to 54 percent (gray whale), lactose ranges from nondetectable (some whales) to 10 percent (green monkey), protein from 1 (zebra) to 20 percent (hare), and ash content from 0.2 (human) to 2 percent (rabbit).

A number of relationships between concentrations of various constituents have been noted in comparisons between species. Milk is isosmotic with blood, but osmotic pressure is maintained by the colloidal properties of milk with the largest contributors to osmotic effects being diffusible ions and carbohydrates. This means for those species with little production of lactose that increased concentrations of ions in milk are necessary to maintain this balance. This likely explains the strong negative correlation between lactose and the sum of sodium and potassium and lactose and chloride

Table 3.1 Gross milk composition of various species

Species	Percentage by weight					
	Water	Fat	Casein	Whey	Lactose	Ash
Human (*Homo sapiens*)	87.1	4.5	0.4	0.5	7.1	0.2
Cow (*Bos taurus*)	87.3	3.9	2.6	0.6	4.6	0.7
Sheep (*Ovis aries*)	82.0	7.2	3.9	0.7	4.8	0.9
Goat (*Capra hircus*)	86.7	4.5	2.6	0.6	4.3	0.8
Horse (*Equus caballus*)	88.8	1.9	1.3	1.2	6.2	0.5
Pig (*Sus scrofa*)	81.2	6.8	2.8	2.0	5.5	1.0
Dog (*Canis familiaris*)	76.4	10.7	5.1	2.3	3.3	1.2
Cat (*Felis catus*)	—	4.8	3.7	3.3	4.8	1.0
Rat (*Rattus norvegicus*)	79.0	10.3	6.4	2.0	2.6	1.3
Mouse (*Mus musculus*)	73.6	13.1	7.0	2.0	3.0	1.3
Blue whale (*Balenopteridae musculus*)	45.5	39.4	7.2	3.7	0.4	1.4
Sperm whale (*Phyeter catodon*)	63.8	25.7	3.8	4.4	0.2	0.6

Source. Adapted from Oftendal, 1984 and 1997, Jenness, 1985.

in milk in comparisons between species. For example, milks of seals and whales contain little carbohydrate but have high concentrations of diffusible ions. In contrast, milk of primates is high in lactose but contains low concentrations of ions. There is also a negative correlation between protein concentration and lactose content of milk between species that is especially evident if expressed on a skim-milk basis. Table 3.1 provides a comparison of gross milk composition for a sampling of species.

Pathways for Protein and Lactose Secretion

In the absence of disease or trauma, most of the proteins present in milk are synthesized from free amino acids or peptides absorbed from the bloodstream. The basolateral cell membrane serves to regulate the uptake of these molecules from the interstitial fluids. Arteriovenous differences in amino acid concentration across the lactating mammary gland are substantial, especially for many of the essential amino acids. Understanding of amino acid transporters is not as well developed for mammary tissue as for other tissues, but many features are likely common between tissues. Some of these transporters show an ion dependence (e.g., Na^+, Cl^-, and K+) or use an H^+-gradient to drive transport. For example, the Na^+ dependent system A transporters for neutral amino acids have been described for lactating mammary tissue, and they are believed to be responsible for the accumulation of some neutral amino acids within the cells when compared with plasma concentrations. There is much interest in regulation of amino acid uptake to better understand factors limiting milk protein synthesis. From a dairy production viewpoint, much of the value of ruminant milk resides in the protein content. Consequently, development of techniques to maintain yields but enhance the protein concentration would be valuable (Bequette et al., 1998).

Classic autoradiographic studies, which traced the movement of radiolabeled amino acids through the alveolar cells, established that the site of milk protein synthesis was the RER. After lactating rats were injected with [^{3}H]-leucine or tissue explants were incubated with a bolus of radiolabeled leucine, the percentage of label in the RER subsequently fell with a following peak in labeling of the Golgi region of the secretory cells. Within 30 minutes of exposure, label began to decrease in the Golgi but increase in the alveolar lumen. These simple but convincing studies demonstrated that after synthesis in the RER, proteins are rapidly transported to the Golgi for packaging into secretory vesicles and subsequent exocytosis. These cytological aspects of protein synthesis and secretion in the mammary gland mimic the patterns of protein secretion by the enzyme-secreting cells of the exocrine pancreas (Heald and Saacke, 1972).

While it is beyond the scope of this book, elegant molecular studies have confirmed that the specific milk proteins are synthesized by membrane-associated ribosomes and that the newly made proteins have short sequences of amino acids that serve as signals to allow binding and vectoring of the nascent protein into the cisternal spaces of the RER. The signal peptide is ultimately cleaved as the protein progresses to the Golgi apparatus for possible posttranslational modification (i.e., enzymatic addition of sugar residues or phosphate groups). The proteins are ultimately released from the Golgi as secretory vesicles. For example, it is generally accepted that individual caseins begin to assume their tertiary structure, aggregate into rudimentary casein micelles, during packaging of secretory vesicles and that mature casein micelles evolve as the vesicles proceed toward exocytosis. In fact, transmission electron microscopy observations suggest that flocculent material present within the Golgi cisternae represents the early stages of this maturation process for micelle formation. Mechanisms responsible for the transport of newly formed secretory vesicles are unknown, but many of these vesicles have associated proteins (i.e., "coated vesicles"). These proteins likely allow interactions with the cytoskeleton and plasma membrane to affect transit and exocytosis of the secretory vesicles.

Enzymes necessary for lactose synthesis are present in the Golgi apparatus. After it is synthesized and its signal peptide directs it into the RER cisternae, the whey protein α-lactalbumin is glycosylated in the Golgi apparatus. During transit it combines with membrane-bound galactosyltransferase to generate the functional enzyme lactose synthase. This enzyme acts to combine the monosaccharides glucose and galactose to produce lactose. With the exception of some marine mammals and during some stages of lactation in marsupials, lactose is the major carbohydrate in milk. Lactose is packaged into secretory vesicles along with other specific milk proteins. However, inclusion of lactose has an important effect on the secretory vesicles. Because it cannot pass across the secretory vesicle membranes, it has a dramatic osmotic effect such that water enters the secretory vesicles. This explains the rather odd appearance of the secretory vesicles of the mammary secretory cells. In the transmission electron microscope, the vesicles appear almost empty except for the dense casein micelles. For other secretory cells that synthesize and secrete proteins, the secretory vesicles often exhibit large, dense protein granules with the membrane-bound vesicles. Figure 3.7 shows a portion of the apical region of a bovine mammary cell from a lactating cow and a portion of a cell from the anterior pituitary gland for comparison. Note the marked difference in the appearance of secretory vesicles in the two cells.

Pathways for Lipid Secretion

Lipid droplets first form as microdroplets near the RER. Through poorly understood mechanisms the droplets progressively enlarge and make their way to the apical plasma membrane. It is theorized that cytoplasm proteins serve to direct this intracellular passage (Mather and Keenan, 1998). These enlarged droplets begin to protrude from the cell, pushing a portion of the plasma membrane into the alveolar lumen. The droplets ultimately pinch off, surrounded by portions of the cell membrane. These membrane droplets are subsequently suspended in the alveolar lumen. Based on transmission electron microscopic examination of mammary tissue of lactating animals or material isolated from milk, the secretory cell loses a small but evident amount of cytoplasm and plasma membrane when each lipid droplet is secreted. In fact in about 1 to 10 percent of globules (depending on species) the cytoplasmic portion included can be substantial. This cytoplasmic piece ranges from a small fragment with a few ribosomes to large fragments of cytoplasm containing rough endoplasmic reticulum, mitochondria or even fragments of Golgi. These so-called cytoplasmic crescents or signets are more frequent in goats, humans, and rats than in cows or sheep (Fig. 3.6). It is likely that this secretory process accounts for at least some of the enzymatic activity measured in milk. Figure 3.6 shows an example of a cytoplasmic fragment from the bovine mammary gland.

Milk fat exists primarily as membrane-bound globules ranging from 0.1 to 15 μm in diameter. The size distribution of globules is an inherited trait and varies both among species and between breeds of cattle. Bovine milk typically contains many small globules that represent only a small fraction of the total milk fat but a large portion of the number of globules. The total number is about 15×10^9/ml with 75 percent of the globules smaller than 1 μm in diameter. Membrane material associated with the fat globules in Holstein-Friesian cows averages about 1.5 g/100 g of milk fat and consists of 60 percent protein and 40 percent phospholipids, along with small amounts of other lipids. The bulk of the lipid within the fat globules (~98 percent) is in the form of triacylglycerols, but small quantities of other lipids are also found as illustrated in Table 3.2.

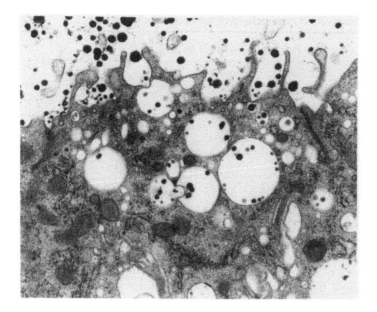

A

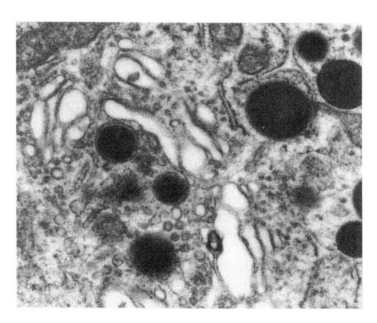

B

Fig. 3.7   Transmission electron microscopy images from the apical region of a secretory alveolar cell from the mammary gland of a cow (**A**) and Golgi region of a bovine pituitary cell (**B**) are shown. Notice that for the mammary cells the secretory vesicles seem to contain space along with scattered casein micelles. This is a reflection of the presence of lactose in the secretory vesicles and associated osmosis of water. In contrast secretory vesicles of the pituitary cells exhibit dense granules with closely adherent membranes. **A** is from Nickerson and Akers, 1984; **B** is an unpublished image (Akers).

Table 3.2 Composition of milk lipid in the cow, human, and rat

Lipid type	Percentage of lipids by weight		
	Cow	Human	Rat
Triglycerides	97–98	98.2	87.5
Diglycerides	0.25–0.48	0.7	2.9
Monoglycerides	0.02–0.04	Trace	0.4
Free fatty acids	0.1–0.4	0.4	3.1
Phospholipids	0.6–1.0	0.25	0.7
Cholesterol	0.2–0.4	0.25	1.6
Cholesterol esters	Trace	Trace	—

Source. Adapted from Jenness, 1985.

The abundant triglycerides make a very diverse population because of the variety of fatty acids. Fatty acid chains range from 4 to 18 carbons with varying degrees of saturation. More that 440 different ones have been identified in cow's milk fat alone. In ruminants because of biohydrogenation in the rumen, nearly all of the dietary fatty acids are saturated before they are available for triglyceride synthesis in the mammary gland. However, unsaturated fatty acids are essential; consequently, the activity of a mammary desaturase enzyme is critical in mammary cell metabolism generally and in milk lipid biosynthesis in particular. Fatty acids in milk fat include those synthesized by the mammary gland (de novo synthesis) as well as those incorporated from dietary lipids. There are variations between orders of mammals, but palmitate (16:0), oleate (18:1), and linoleate (18:2) are often among the most abundant fatty acids. Butyrate (4:0) appears in ruminant milk but is rare in milk of other animals. The distribution of fatty acids within the triglyceride is not random and differs between species. In ruminants butyrate is usually located in the sn-3 position, palmitate occurs equally between the sn-1 and sn-2 position, and oleate occurs at the same rate in all three positions.

Recent studies have focused on a family of geometric and positional isomers of linoleic acid called *conjugated linoleic acids* (CLAs). CLAs have been shown to have potent health benefits including inhibition of carcinogenesis, reduction in atherogenesis, and prevention of diabetes (Parodi, 1999). Milk and dairy products are the primary sources of CLAs in the human diet. The major isomer in bovine milk is cis-9, trans-11, although a number of other isomers are present. CLAs are intermediates in rumen microbial biohydrogenation of unsaturated fatty acids, and these are believed to be the source of these molecules in milk. Manipulation of rumen fermentation by feeding cows plant oils with unsaturated fatty acids can enhance the content of CLAs in milk fat, as does direct dietary supplementation with purified CLAs. Cows on pasture also have higher milk fat CLAs (Kelly et al., 1998). A comparison of milk fatty acid composition is given in Table 3.3.

Table 3.3   Fatty acid composition of milk fat of cow, human, and rat

| Fatty acid | Percentage of fatty acids by weight | | |
	Cow	Human	Rat
4:0	4.0		
6:0	2.6		
8:0	1.4		1.1
10.0	3.1	1.3	7.0
12.0	3.3	3.1	7.5
14:0	10.6	5.1	8.2
15:0	0.9	0.4	
16:0	27.7	20.2	22.6
16:1	0.92	5.7	1.9
17:0	0.5		0.3
18:0	8.2	5.9	6.5
18:1	15.4	46.4	26.7
18:2	2.4	13.0	16.3
CLAs	0.47		
18:3	0.8	1.4	0.8

Source. Adapted from Jenness, 1985, and Kelly et al., 1998.

In addition to the triglycerides, there are also small amounts of mono- and diglycerides and free fatty acids. The phospholipids made up only a small portion of the total lipids, but these molecules make a complex mixture. Phosphoglycerides are derived from phosphatidic acid, which consists of glycerol with fatty acids esterified in positions 1 and 2 and phosphoric acid in the third position. Four variants include phosphatidylcholine (PC), phosphatidylethanolamine (PE), phosphatidylserine (PS), and phosphatidylinositol (PI), in which the indicated compound is esterified to the phosphate. Variants of sphingomyelin (SPH) make up the bulk of the remaining milk phospholipids. Proportions of these molecules are similar between various species.

Secretion of Other Milk Constituents

In addition to mammary cell–specific constituents, milk contains a myriad of minor components. Many of these molecules are important nutrients or regulators of the neonate (growth factors, water, ions), but other components may include drugs or other xenobiotic substances transported from the circulation. Table 3.4 compares the average mineral content of bovine and human milk.

Molecules are transported into the milk by several possible routes. Like most cells the mammary epithelial cells are able to maintain substantial gradients for Na^+, K^+, and Cl^- ions across the cell membrane. During established lactation there

Table 3.4 Mineral content of bovine and human milk

Mineral	Bovine (mg/dl)	Human (mg/dl)
Calcium	125	33
Magnesium	12	4
Sodium	58	15
Potassium	138	55
Chloride	103	43
Phosphorus	96	15
Citrate	175	60
Sulfur	30	14

Source. Adapted from Jenness, 1974.

are also gradients between milk and plasma. These ions are important certainly to maintain normal electrical gradients across the alveolar cell membranes, but they also are critical in regulation of milk osmolarity, especially for those species with low lactose production. Concentrations of Na^+ inside (~43 mM) the cells are typically lower than outside (150 mM), but the gradient for K^+ is the opposite (143 mM inside compared with 4.5 mM). These differentials are maintained by action of Na^+ K^+ ATPase pumps in the basolateral membranes. The apical plasma membrane is permeable to both ions so that the distribution of these ions into milk is controlled by the electrical potential across the apical plasma membrane. Milk is electrically positive with respect to the cell so that the concentrations of Na^+ and K^+ are lower in milk than in the cells, but the K^+/Na^+ ratio (~3:1) is similar. Concentrations of Cl^- are higher inside the cells than the equilibrium distribution would suggest, so there are likely membrane pumps in the basolateral and apical membranes that act to sequester Cl^-. It is easy to imagine that this balance of ions between milk and blood is readily compromised if the leakiness of the tight junctions is altered (Stelwagen, 2001).

As dietitians and nutritionists frequently note, milk is a rich source of calcium, with total concentrations equaling 100 mM or more. With the onset of lactation, the mammary gland extracts large quantities of calcium to supply the developing neonate. Indeed, for high-producing dairy cows, this demand can lead to metabolic periparturient paresis unless animals are carefully managed. The calcium in the milk exists as free calcium, casein-bound calcium, or calcium associated with various inorganic anions (e.g., citrate and phosphate). There is apparently little movement of calcium from milk to blood, suggesting that calcium cannot pass across the apical plasma membrane of the tight junctions. Given that most of the calcium is associated with the casein micelles, the Golgi vesicle route of secretion is the predominating pathway. However, since all of the milk calcium is derived from the circulation, there must be differences in transport between basolateral and apical membranes. Mammary cells main-

tain a low intracellular free calcium concentration in spite of the marked accumulation of calcium in milk. This is important given that changes in free calcium concentration are closely linked with several hormone and growth
factor–signaling pathways. One idea is that the rate of calcium influx into the
cell is matched by a corresponding uptake of calcium by cellular organelles. The
presence of an ATP-dependent calcium pump on Golgi membranes has been
demonstrated. Calcium uptake by mammary cells may also be hormonally regulated since parathyroid hormone–related protein and $1,25\text{-}(OH)_2$ vitamin D_3
stimulate the uptake of calcium in cultured mammary tissue.

More that 50 enzyme activities have been detected in bovine milk and many in
the milks of other species. Some of these proteins are derived from the bloodstream or from leucocytes, but most are derived from the mammary cells. Most of
these enzymes appear in milk as a consequence of normal secretion. However,
concentrations of and activities of milk enzymes vary between species and
between individual animals. Activities are generally greater during colostrum formation than during established lactation. Also mastitis or other processes that
increase leucocytes in the milk increase enzyme activity. Activity and/or concentrations of the enzyme plasmin are increased in late lactation. This enzyme acts to
degrade casein and thereby make late lactation milk less suitable for cheese making. This can be an especially troubling problem for milk product manufacturing
in areas of the world with marked seasonal breeding of dairy cows since large
numbers of animals are in late lactation simultaneously.

Milk contains a large number of compounds that occur in low concentrations (i.e.,
less than 100 mg/L). These include various metabolites including nucleotides, hormones, vitamins, alcohols, ketones, and gases. Numerous trace elements are also normally present in milk. Molybdenum (Mo) is associated with the protein xanthine
oxidase and cobalt (Co) as part of vitamin B_{12}. Iron (Fe) appears as a component of
xanthine oxidase, lactoperoxidase, and catalase. Among metals zinc (Zn) is at the
highest concentration and is associated with casein micelles or citrate. Copper (Cu)
and Fe may also be associated with the fat globule membrane. Other trace elements
may enter milk by contamination after milking. Metal containers can be a source of
Cu, Fe, nickel (Ni), or tin (Sn). Concentrations of some of these elements may also
be modified by changes in the diet.

Nonprotein nitrogen (NPN) compounds have low molecular weights and are
not precipitated by 12 percent trichloroacetic acid. The total for these compounds
varies considerably between species. Bovine milk averages about 300 mg/L. This
equals about 6 percent of total milk nitrogen. Human milk averages about 400
mg/L, but the differences in average protein content mean that these compounds
account for about 17 percent of milk nitrogen. For both species urea accounts for
about half of the total milk NPN. Recent studies have touted measurement of milk
urea nitrogen (MUN) as a tool for dairy nutritionists to evaluate protein metabolism. Free amino acids and small peptides also contribute to milk NPN in bovine
milk. Other contributors to milk NPN include ammonia, creatine, creatinine, uric
acid, and orotic acid (particularly in bovine milk).

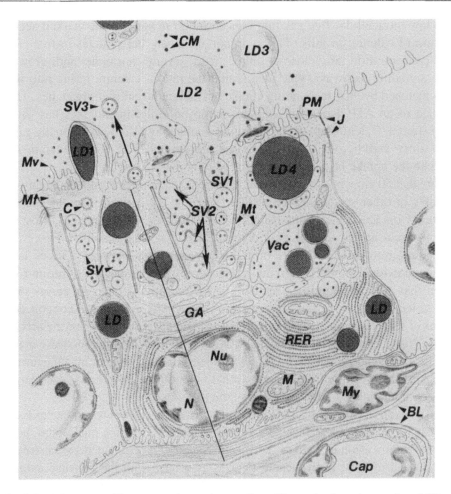

Fig. 3.8 Diagram to illustrate major pathways for milk synthesis and secretion. Milk precursors in capillaries (Cap) are transported across the endothelial cells and basal lamina (BL) to the interalveolar connective tissue. Nutrients pass across the alveolar BL and/or myoepithelium (My), over the basal plasma membrane (PM), and into the cytoplasm. Milk proteins are synthesized in the rough endoplasmic reticulum (RER), enter the RER lumena, and are transported to the Golgi (GA) for processing and packaging. In typical exocytosis the secretory vesicles (SV) with casein micelles (CM) and lactose leave the Golgi, translocate to the apical PM, and release contents of the vesicle (SV1). Alternatively, vesicles can fuse to form chains for secretion (SV2) or fuse with release of double membrane-bound micelles (SV3). Milk lipid is synthesized in the region of the RER, and as droplets grow, they also translocate to the apical PM. These lipid droplets (LD) are enveloped by PM, protrude from the cell (LD1), are pinched off (LD2), and then secreted into the lumen (LD3). It is also possible that SV can fuse around lipid droplets, with other droplets, and with the apical PM in groups (LD4). Lipid droplets might also be released via coalesced secretion vacuoles (Vac). Other features include mitochondria (M), nucleus (N), nucleolus (Nu), microtubules (Mt), microfilaments (Mf), microvilli (Mv), coated vesicles (C), and tight junctions (J). Diagram modified from Nickerson and Akers, 1984.

Contaminants may also appear in milk. These include materials that are not normally secreted into milk but appear either by accident or design. Some of these substances are eaten or inhaled by the lactating animal and pass into milk via the bloodstream and mammary gland. Examples include various chemicals (i.e., pesticides and herbicides). An unfortunate example in the late 1970s occurred when polychlorinated biphenyls (PCBs) were accidentally included in feed supplements. Because they were lipophilic, they accumulated in fat depots and were secreted in association with milk fat. Other materials can enter milk after animals are treated by intramammary infusions. The appearance of antibiotics in milk fits this example because it is simply flushed into milk. Antibiotic residues can cause problems with individual consumers and negatively impact cheese making. Finally, some contaminants enter milk during milking, processing, or storage. Figure 3.8 provides a diagram showing the primary secretion pathways for a fully differentiated alveolar cell. Although the secretion process is generally well understood and adequately described, some major gaps remain. For example, precise mechanisms for intracellular lipid droplet growth and translocation are unknown.

Chapter 4
Milk Component Biosynthesis

An understanding of milk synthesis depends on an appreciation of the metabolic pathways necessary to supply the precursors needed to produce the protein, fat, and carbohydrate in milk in addition to understanding synthesis of these specific milk components. The primary substrates extracted from blood by the lactating mammary gland are glucose, amino acids, fatty acids, and minerals. For ruminants because of bacterial fermentation of dietary carbohydrates, acetate and β-hydroxybutyrate are also critical substrates. Glucose is the direct precursor for lactose, ribose, and much of the glycerol needed for triglyceride synthesis. Energy requirements in the form of ATP are satisfied by the oxidation of glucose and/or fatty acids in nonruminants but largely by acetate in the ruminant. All the essential amino acids and many nonessential amino acids are derived from the bloodstream. The purine and pyrimidine nucleotides needed for synthesis of RNA and DNA and lactose synthesis are produced by the epithelium. Fatty acids for production of triglycerides are derived from the bloodstream (i.e., dietary lipids), mobilization from body stores, as well as de novo synthesis by the alveolar cells. The coenzymes (e.g., nicotinamide dinucleotide phosphate [NADPH]) necessary for de novo fatty acid synthesis or for ribose synthesis are derived by oxidation of isocitrate in the Krebs cycle or the pentose phosphate shunt.

Essential metabolic events and associated cellular organelles necessary for milk biosynthesis are outlined in Figure 4.1. First, transport processes at the level of the cell membrane, especially the lateral-basal membranes, are critical because the precursors for milk synthesis are derived from the bloodstream. The nucleus provides the genetic blueprints, the transcription of DNA, to make the mRNA that is subsequently translated by the ribosomes to create the myriad of enzymes necessary for milk biosynthesis. The rough endoplasmic reticulum is also the site for esterification of fatty acids to glycerol and ultimately lipid droplets production. This is in addition to generation of phospholipids and desaturation of fatty acids (especially relevant in ruminants). The mitochondria, whose numbers increase with the onset of lactation, are the sites for production of ATP, generation of precursors for synthesis of nonessential amino acids, and formation of building blocks for de novo fatty acid synthesis. The Golgi apparatus is where lactose is

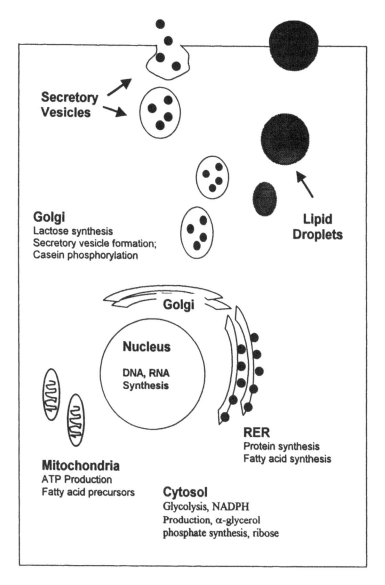

Fig. 4.1 Structure-function relationships in mammary epithelial cells. Adapted from Davis and Bauman, 1974.

produced and together with milk proteins packaged into vesicles for secretion. Other metabolic processes in the Golgi include casein phosphorylation, maturation of the casein micelles, and glycoprotein production. The cytoplasm is the site for glycolysis, production of α-glycerol phosphate, and amino acid activation for protein synthesis. Thus, the fully functional mammary secretory cell can be viewed as functionally polarized with the basal-lateral region occupied with uptake of precursors and biosynthesis, the Golgi with final manufacturing and packaging, and the apical region of the cell with secretion.

Metabolic Pathways and Precursors for Milk Biosynthesis

Understanding milk biosynthesis requires knowledge of pathways to produce ATP and reduced nicotinamide adenine dinucleotide (NADH), the biological significance of substrate versus oxidation phosphorylation, anaerobic versus aerobic respiration, glycolysis, the Krebs cycle, the electron transport chain, and the pentose phosphate shunt. An eager student is also well served by having a grasp of translation and transcription, oxidation deamination and transamination of nonessential amino acids, and β-oxidation of fatty acids.

Glycolysis, Gluconeogenesis, Krebs Cycle, and Energy Requirements

The most important substrates for production of energy in mammary cells are glucose and acetate. Use of these substrates for energy production, however, are markedly different in ruminants and nonruminants. Nonruminants have relatively higher circulating glucose concentrations but usually little acetate. So for these animals glucose is readily used for generation of ATP as well as lactose synthesis. Because of fermentation of dietary carbohydrates, ruminants absorb little glucose from the small intestine. Fortunately, the volatile fatty acids derived from fermentation provide acetate for generation of ATP and propionate (largely by the liver) to generate a sufficient quantity of blood glucose. Because maintenance of blood glucose depends on gluconeogenesis, in ruminants the mammary cells minimize use of glucose for production of energy. However, with the onset of lactation, a large portion of blood glucose is ultimately diverted for lactose synthesis; thus, demands on gluconeogenesis are markedly increased. Failure to make the biochemical adjustments to meet these demands can lead to a variety of metabolic disorders including milk fever, ketosis, and fatty liver disease.

Whether glucose is destined for lactose synthesis, fat synthesis, or generation of ATP, glycolysis forms an essential framework for understanding milk synthesis. Following digestion of complex carbohydrates or more simple sugars, glucose enters the mammary cell. If it is not directly destined for lactose synthesis (described in a later section), the glucose is phosphorylated to form glucose-6-phosphate by the enzyme hexose kinase. This effectively "traps" the glucose in the cell because the addition of the phosphate group minimized interaction with membrane glucose transporters. At this point the glucose can be utilized in one of three pathways. These are (1) glycolysis to produce pyruvate and a limited amount of ATP, (2) formation of galactose (necessary for lactose production), or (3) formation of pentose phosphates (e.g., ribose via the pentose phosphate cycle). Although it is routine to study these various biochemical pathways independently, it is important to appreciate that they are closely interrelated. For example, the pentose phosphate pathway is intertwined with pathways for fatty acid synthesis because production of the reduced form of the cofactor nicotinamide dinucleotide phosphate (NADPH) is essential to provide the hydrogen ions needed for fatty acid synthesis. Activity of the fatty acid synthesis pathway, in turn, generates

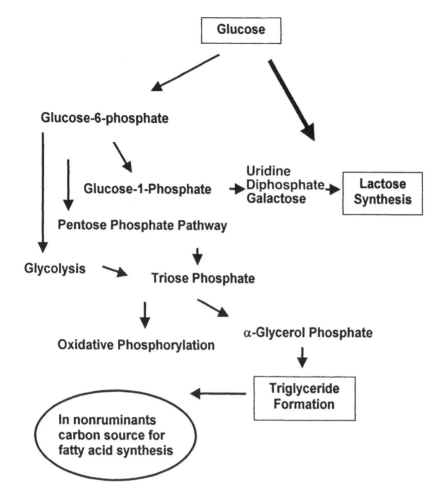

Fig. 4.2 Pathways for glucose metabolism in mammary epithelial cells.

reduced NADP used in the pentose phosphate shunt. Other intermediates of the pentose phosphate pathway are used to synthesize nucleic acids. Figure 4.2 outlines the uses of glucose in mammary cells from a lactating animal.

Conversion of the hexose sugar glucose to pyruvate in glycolysis yields two molecules of the interconvertible triose sugar glyceraldehyde phosphate along the way. If it is subsequently oxidized, the removed hydrogen serves to generate NADH. Inorganic phosphate groups added to the oxidized sugars are ultimately cleaved to provide energy for generation of a total of 4 ATP (i.e., so-called substrate-level phosphorylation). This explains the low yield of ATP derived from anaerobic respiration. The final products of glycolysis are two molecules of pyruvate, which in the presence of sufficient oxygen are shunted through the Krebs cycle. Briefly, pyruvate is converted to acetyl coenzyme A. This involves (1) decarboxylation in which one of the carbon atoms of pyruvate is removed and released as CO_2, (2) oxidation by removal of hydrogen atoms (i.e., conversion of NAD to NADH), and (3) combining the resulting acetic acid with coenzyme A to produce acetyl coenzyme A, which is often referred to as simply acetyl CoA. Essentially, coenzyme A shuttles the two-carbon acetic acid to be enzymatically combined with the four-carbon intermediate oxaloacetic acid. The bond between the acetyl group and CoA is broken as the bond with oxaloacetic acid is created. This frees the CoA to prime another two-carbon unit from pyruvate. What is formed from this reaction is the six-carbon molecule citric acid. Indeed, the Krebs cycle is often also called the citric acid cycle. A relatively small amount of additional ATP is generated by substrate-level phosphorylation (i.e., as succinyl-CoA is converted to succinic acid). More important for the sake of ATP production is the fact that for each turn of the Krebs cycle three molecules of reduced NADH and one of reduced flavin adenine dinucleotide (FADH) are generated. Transfer of these reduced coenzymes to the electron transport chain in the mitochondria coupled with oxidation of the molecules allows the creation of a proton gradient between the inner and outer membranes of the mitochondrial matrix. This gradient provides the energy ultimately captured by the enzyme ATP synthase to generate ATP from ADP and inorganic phosphorus (P_i). This is the essence of oxidative phosphorylation. For the sake of perspective, it should be noted that the anaerobic substrate phosphorylation reactions of glycolysis yield only a net gain of 2 ATP per molecule of glucose metabolized but a total of 36 in the presence of oxygen and the coupled reactions of the Krebs cycle and the electron transport chain. It is also worth pointing out that CoA is derived from pantothenic acid or vitamin B_5. Similarly, vitamin B_2 (riboflavin) and niacin are essential for creation of oxidized flavin adenine dinucleotide (FADH) and NADH, respectively.

However, glyceraldehyde phosphate is not only a critical intermediate in the pentose phosphate shunt pathway but also the major precursor for production of α-glycerol phosphate. Triglyceride synthesis uses α-glycerol phosphate as the carbon backbone for attachment of fatty acids. In a related fashion some of the available glyceraldehyde phosphate that continues toward pyruvate production results in the production of reduced NADH, which provides the reducing power necessary for α-glycerol phosphate production or for production of malate in the malate transhydrogenation cycle. However, ruminants and nonruminants are very different in utilization of the NADH generated by pyruvate production from glyceraldehyde phosphate. Specifically, activity of the malate transhydrogenation cycle is markedly reduced in ruminants. This means that little of the citrate, derived from the condensation of pyruvate and oxaloacetate in the Krebs cycle, is used to provide the acetyl CoA needed for fatty acid synthesis. Second, acetate, which is abundant in ruminate blood, easily enters the mitochondria to provide acetyl CoA for generation of ATP. In effect this serves to spare utilization of glucose for ATP production or to generate the carbon skeleton needed for de novo fatty acid synthesis. Given the premium placed on use of glucose in lactose production, this is an important adaptation to minimize metabolic stress in high-producing cows. Another of the primary volatile fatty acids derived from microbial fermentation, β-hydroxybutyric acid (BHBA) is mostly used by mammary

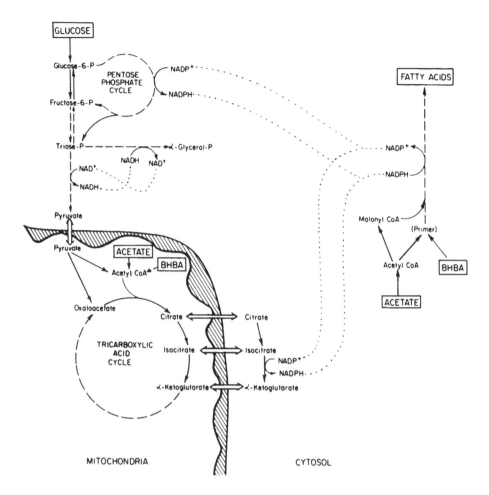

Fig. 4.3   Biochemical pathways related to fatty acid synthesis in the ruminant mammary gland. Adapted from Bauman and Davis, 1974.

cells in synthesis of fatty acids. The primary reason that little acetate is used for ATP generation or for fatty acid synthesis in nonruminants is simply its low concentration in blood. Figures 4.3 and 4.4 illustrate the interrelationship of the pathways for fatty acid synthesis in ruminants and nonruminants.

Lactose Biosynthesis

The disaccharide lactose is the most common carbohydrate found in milk. It is one molecule of glucose and one of galactose combined in a one to four carbon linkage as a β-galactoside. Its concentration ranges from a trace in many aquatic mammals to 7 percent in primates. The functional enzyme required for lactose synthesis, lactose synthase, is actually a combination of two proteins that come

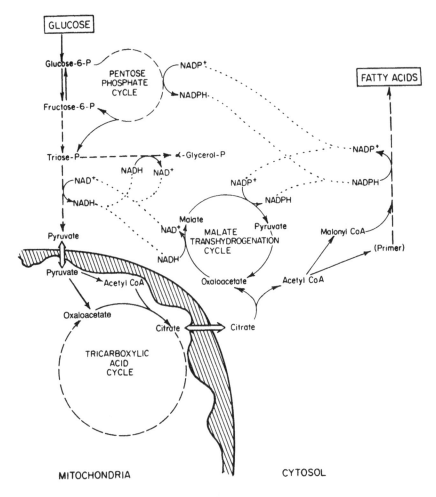

Fig. 4.4 Biochemical pathways for fatty acid synthesis in the nonruminant mammary gland. Adapted from Bauman and Davis, 1974.

together in the Golgi apparatus. The pathway for synthesis involves three general steps:

(1) UTP + glucose   → UDP-glucose + P-P
(2) UDP-glucose   → UDP-galactose
(3) UDP-galactose + glucose → lactose + UDP

where UTP = uridine triphosphate, UDP = uridine diphosphate, UDP-glucose = uridine diphosphoryl glucose, UDP-galactose = uridine diphosphoryl galactose, and P-P = pyrophospate. The enzymes involved in these steps are (1) uridine diphosphoryl glucose pyrophosphorylase, (2) uridine diphosphoryl galactose-4-epimerase, and (3) lactose synthase (a combination of galactosyl transferase and a-lactalbumin).

UTP is derived from the pyrimidine synthesis pathway. UDP galactose-4-epimerase in mammary tissue appears well before parturition, but its activity is increased near the onset of lactation. However, this enzyme is also present in other tissues. In contrast, the activity of lactose synthase is markedly increased in close association with the final stages of lactogenesis. One part of the functional enzyme, the galactosyl transferase, is widely distributed and acts to transfer galactosyl groups from UDP-galactose to the *N*-acetylglucosamine residues present in the sugar moieties of many general cell glycoproteins. The ability to make lactose depends on the appearance of the protein α-lactalbumin in the mammary cells. Specifically, only in the presence of α-lactalbumin does the galactosyl transferase acquire the capacity to transfer a galactosyl group to free glucose, thereby creating lactose. This is an extreme example of allosteric modulation of enzyme activity. It is generally believed many cells routinely synthesize galactosyl transferase and that it is subsequently bound on the inner surfaces of the Golgi membranes where it functions in the routine production of glycoproteins. As expression of the α-lactalbumin gene increases with lactogenesis, the capacity of the mammary cells for lactose synthesis is subsequently activated as newly synthesized α-lactalbumin transits through the Golgi apparatus and interacts with the membrane-bound galactosyl transferase. Because the α-lactalbumin ultimately is secreted into milk, lactose production requires the continuous production of α-lactalbumin to generate the functional lactose synthase enzyme complex.

Milk Fat Biosynthesis

As previously illustrated, the fat content of milk shows very marked differences between species, from less than 1 percent to more that 50 percent. Even within species, fat content is the most variable of the major milk constituents. For example, the fat content of the milk of dairy cows is readily altered by changes in diet. Most of the milk fat is composed of triglycerides, but particular fatty acids are highly variable. This variety is largely a reflection of species differences and dietary sources of fatty acids. The three sources of the fatty acids in milk triglycerides are (1) glucose via conversion to pyruvate, citrate, and ultimately acetyl CoA in the cytoplasm, (2) diet via hydrolysis of chylomicra, and (3) de novo synthesis within the mammary cells from nonglucose sources. Glucose is a major source for nonruminants but not for ruminants.

Estimates for cows are that half of the milk fatty acids are derived from the diet, including most of the C_{18} fatty acids and about 30 percent of the C_{16} fatty acids. Generally, dietary fatty acids are greater than 14 carbons and mostly C_{16} (palmitic) and C_{18} fatty acids (stearic, oleic, or linoleic). Shorter-chain fatty acids are more likely to derive from de novo synthesis. In cows precursors for de novo fatty acids synthesis are acetate and BHBA. BHBA appears in the first four carbons of the majority of fatty acids made in the cells, or the molecule is cleaved into two carbon units to be used as acetyl CoA. Acetate yields the carbon for the shorter fatty acids (C_4–C_{14}) and some C_{16} fatty acids. The NADPH needed comes either from the catabolism of glucose via the pentose phosphate shunt or the oxidation of isocitrate to α-ketoglutaric acid in the Krebs cycle. The malonyl CoA pathway, which sequentially adds two carbon units to the growing fatty acid chain, is the major synthesis pathway in the ruminant mammary gland and occurs in the cytoplasm. The first step depends on the regulatory enzyme acetyl CoA carboxylase and involves the addition of carbon from CO_2 to acetyl CoA and hydrolysis of ATP to form malonyl CoA. The second step is catalyzed by fatty acid synthase. This complex enzyme controls growth of the growing fatty acid chain two carbons at a time. Interestingly, in most fat-synthesizing tissues, fatty acid synthase produces mostly palmitic acid. However, the presence of the enzyme thioesterase II in mammary tissue induces the synthesis of more medium-chain fatty acids and fewer long-chain fatty acids. Differences between fatty acid synthesis in ruminants and nonruminants concern the sources of the acetyl CoA needed in the initial step and generation the necessary NADPH.

In nonruminants the acetyl CoA comes from the decarboxylation of pyruvate in the mitochondria but not directly. Specifically, since acetyl CoA does not easily pass the mitochondrial membrane, citrate derived from the combination of acetyl CoA and oxaloacetate diffuses from the mitochondria and enters the cytoplasm. The citrate is broken down by ATP-citrate lyase to give acetyl CoA and oxaloacetate. The acetyl CoA provides carbon for fatty acid synthesis, and the oxaloacetate enters the malate transhydrogenation cycle, which yields pyruvate and NADPH. The pyruvate subsequently enters the mitochondria.

In ruminants acetate and BHBA from the blood provide most of the carbon needed for fatty acid synthesis. Glucose is largely spared from being used as a carbon source because of the near absence of the citrate lyase enzyme in the cytoplasm of the cell. The citrate that does leave the mitochondria is either converted to isocitrate and then α-ketoglutarate, generating NADPH in the process, or it passes into the Golgi and is secreted into milk. Interestingly, cow's milk is higher in citrate than nonruminants', and concentrations of citrate increase with the final stages of lactogenesis. Indeed, this abrupt increase in citrate concentrations of mammary secretions can be used as a marker for lactogenesis and parturition.

Milk Protein Biosynthesis

Although the protein in milk can arise from several sources, immunoglobulins and other proteins transported from the bloodstream, or proteins derived from apocrine secretion, the focus of this discussion is on milk-specific proteins. These proteins are synthesized from amino acids derived either from the bloodstream (essential amino acids) or from amino acids synthesized by the secretory cells. These mammary-specific proteins include the caseins (α_s-casein, β-casein, κ-casein, and γ-casein) and the whey proteins (α-lactalbumin and β-lactoglobulin). The caseins in cow's milk are empirically defined as being those proteins that are precipitated when skim milk is acidified to pH 4.6 at 20°C. They also account for about 80 percent of the specific milk proteins. Compared with other common proteins, the caseins are rich in proline and glutamic acid, with less than average amounts of glycine and asparagine. The caseins are also relatively hydrophobic, and under the usual ionic conditions in milk, they strongly associate to form the familiar casein micelles. In electron microscopic images the micelles are generally spherical with a granular appearance, and most range from 10 to 300 nm in diameter. The micelles are not only a nutritious source of amino acids but the colloidal structure of the micelle also allows the transport of large amounts of calcium and phosphorus in a stable form. The micelle is about 90 percent casein and is also made up of calcium, inorganic phosphorus, and small amounts of citrate and magnesium. The α-s$_{1-}$, α-s$_{2-}$, β- and κ-caseins appear in a ratio of approximately 3:1:3:1, but there are variations since the proportion of κ-casein is increased in smaller micelles. The κ-caseins are glycosylated and also differ from the other caseins because they are easily degraded by the enzyme chymosin (rennin) to yield to fragments called para-κ-casein and glycomacropeptide. This property accounts for the ability of rennin to rapidly coagulate milk and illustrates the critical importance of κ-caseins in stabilizing the casein micelle. In summary, the α-s$_1$ and α-s$_2$ caseins account for about 49 percent, β-casein another 35 percent, and κ-caseins about 10 percent of the total caseins in milk. Small quantities of three variants of γ-caseins are produced by the proteolytic action of the enzyme plasmin against β-casein, which accounts for most of the other casein fragments in milk. Although proportions vary, the mature milk of all known species contains similar proteins.

The whey or milk serum proteins, which remain in solution when milk is acidified to pH 4.6, contain a variety of proteins. These include β-lactoglobulins, α-lactalbumin, immunoglobulins, serum albumin, lactoferrin, and transferrin. Some of these are mammary-specific proteins, but some (e.g., immunoglobulin and albumin) are derived from blood. Distribution varies between species. For example, α-lactalbumin is very widely distributed, but β-lactoglobulin is present in ruminants but essentially absent in the milk of humans, rats, and guinea pigs. It is believed that many of the milk protein allergies associated with feeding bovine milk to humans are especially related to β-lactoglobulin. β-Lactoglobulin is the major whey protein in the milk of cows; it accounts for more that 50 percent of the whey protein. At least five genetic variants have been identified in cow's milk as well as two each in the milk of sheep and goats. The protein is relevant nutritionally but has no other identified biological role. In contrast, α-lactalbumin is the second most abundant whey protein and, as discussed in the section on lactose biosynthesis, is essential for the mammary cell to synthesize lactose.

Lactoferrin and transferrin are glycoproteins, which transport and bind iron. Lactoferrin especially is bacteriostatic and is believed to be important as a nonspecific component in defense of the mammary gland. In human milk lactoferrin is only about 8 percent saturated with iron. Availability of unsaturated lactoferrin serves to inhibit the growth of Gram-positive and Gram-negative bacteria in secretions because of its ability to sequester iron and limit its use by the bacteria. Mature milk of different species varies in the relative content of lactoferrin and transferrin. Human milk is high in lactoferrin (1–6 mg/ml) but has little transferrin. Bovine milk is low in lactoferrin, with trace amounts of transferrin derived from the plasma. Rabbits and rats have transferrin of mammary origin but little or no lactoferrin, whereas dogs have little of either protein. In both cows (20–100 mg/ml) and humans (~50 mg/ml), concentrations of lactoferrin are greatly increased in mammary secretions obtained when the mammary glands are involuting. This is especially fitting for cows since the dry period between lactations, just before calving, is the time when the animals are most susceptible to mastitis. Interestingly, this pattern of increased synthesis and secretion of lactoferrin in the involuting mammary gland just when secretion of the usual milk proteins wanes is paradoxical but has been confirmed at the mRNA level. Moreover, immunosustaining and in situ hybridization studies with labeled RNA probes showed that lactoferrin mRNA and protein were primarily expressed in localized alveoli, with minimal expression of α-lactalbumin and casein.

Protein synthesis in mammary cells follows the pattern described for many other tissues. Aside from directing its own replication, DNA also serves direct protein synthesis by its capacity to generate mRNA. Generally, each gene is composed of a segment of DNA, which carries the chemical instructions for synthesis of one polypeptide chain in its arrangement of nucleotide bases (adenine, thymine, cytosine, and guanine). Each sequence of three bases, a triplet code, acts to direct the joining of a specific amino acid in the mature mRNA molecule. Although one-half of the double-stranded DNA serves as template for synthesis of the mRNA (transcription), not all of the nucleotides in the gene appear in the final mRNA blueprint. The genes of higher organisms contain exons, the amino acid–specifying sequences, separated by introns. These noncoding introns range from 60 to 100,000 nucleotides in length. Transcription of a particular gene depends on the binding of a transcription factor to a site on the DNA adjacent to the start sequence for the gene. This region is the promoter. The transcription factor mediates the binding of the enzyme RNA polymerase. This enzyme acts to open the DNA helix, and the DNA segment coding for the protein is uncoiled. Only one strand of the DNA, the sense strand, serves as the template for creation of a complementary mRNA. However, before the mRNA can direct protein synthesis, the noncoding introns are enzymatically removed before the newly made mRNA exits the nucleus for translation. Single-stranded RNA also differs from double-stranded DNA in having the sugar ribose instead of deoxyribose and the base uracil instead of thymine. These features provide a ready means to access the ability of mammary cells to proliferate or synthesize proteins by measuring the incorporation of radiolabeled thymidine or uracil.

Protein synthesis depends on three forms of RNA. These are (1) transfer RNA, (2) ribosomal RNA, and (3) messenger RNA. When the mature mRNA reaches the cytoplasm, it binds to a small ribosomal subunit by base pairing to ribosomal RNA (rRNA). The transfer RNA (tRNA) has the task of transferring amino acids to the ribosome. There are approximately 20 different types of tRNA, each capable of binding a specific amino acid. The linkage process is controlled by a synthetase enzyme and depends on the cleavage of ATP to form the peptide bonds between amino acids of the growing peptide chain. Once its amino acid is loaded, the tRNA migrates to the ribosome where it moves the amino acid into position based on the codons of the mRNA strand. The amino acid is bound to one end of the tRNA (the tail), but the other end of the molecule (the head) has a three-nucleotide base sequence (anticodon), which is complementary to the codon of the mRNA. For a given strand of mRNA, multiple ribosomes can become attached, and as the ribosomes move along the molecule, many chains of new protein can be made simultaneously. In fact, it is not uncommon to find polyribosomes in the cytoplasm. These are represented in transmission electron microscopic views of mammary cells by chains or coils of ribosomes seemingly organized in the cytoplasm.

However, as mentioned, proteins destined for secretion from the cell are synthesized by ribosomes attached to the endoplasmic reticulum. The mRNA for these proteins codes an initial short peptide sequence (signal peptide), which directs the growing peptide chains into the cisternal space of the endoplasmic reticulum. Because this space is continuous with the Golgi apparatus, proteins destined for secretion are vectored into Golgi for packaging into secretory vesicles and secretion from the cell by exocytosis. After synthesis in the rough endoplasmic reticulum (RER), modifications to secretory proteins may occur in the Golgi apparatus as well. These posttranslational modifications can markedly affect the function of the protein. For example, in the presence of calcium and inorganic phosphate in the Golgi, the casein molecules aggregate into micelles. But this does not occur until the individual caseins have been glycosylated and phosphorylated and ion composition of the Golgi allows formation of the final three-dimensional structure of the complex micelle as it appears in milk.

Biotechnologies and the Mammary Gland

One of the truly exciting applications of biotechnology has been the realization in recent years that the capacity of the mammary gland to synthesize and secrete proteins could be harnessed to produce novel, therapeutically important proteins. Use of isolated, purified proteins in medicine is nothing new. A variety of plasma-derived products have been used for many years. These include immunoglobulin preparations, serum albumin, and various enriched blood-clotting fractions. However, with the exception of some of the blood-derived products and insulin isolated from porcine or bovine pancreatic tissue, use of many proteins with potential health or industrial value was limited by the lack of stable supplies or available resources for isolation. For example, until the advent of recombinant DNA technology, the only sources of human growth hormone was from human cadavers. In agriculture, use of recombinant bovine growth hormone (bST) to increase milk production in lactating dairy cows has been a great success. Similarly, bacterial cells now are utilized to produce virtually unlimited supplies of recombinant human growth hormone or insulin. Because insulin and the growth hormones are simple proteins that do not require posttranslational modifications for bioactivity, production in bacterial cultures is successful. However, many proteins with potential therapeutic value are highly modified (i.e., carboxylated, glycosylated, or otherwise modified) prior to secretion. This processing requires eukarotic cells. Although culture of mammalian cells on a commercial scale is possible, the task is daunting and expensive. Perhaps more importantly, few cells in culture are capable of synthesizing large quantities of proteins. The search for a suitable production method ultimately led to use of the mammary gland as a bioreactor.

Initial studies in mice described the production of transgenic mice carrying copies of the rat growth hormone gene. In these studies the cloned rat gene was introduced into the germline by physical microinjection of the DNA into the pronuclei of fertilized mouse eggs. These injected eggs were returned to suitable donors, a portion of the eggs produced offspring, and a number of the animals expressed the foreign gene in their tissues. Because the gene construct had been fused with a portion of the promoter region of the metallothionein gene, large amounts of rat growth hormone were synthesized by the liver of the animals and secreted into blood. Transgenic animals grew more rapidly and reached a greater size than controls. Such studies demonstrated the feasibility for targeting expression of a particular foreign protein to a specific tissue. Despite the logistical and practical hurdles, since this time microinjection techniques have yielded transgenic animals for all the major commercial livestock species, including sheep, goats, pigs, and cows (Clark, 1998; Ayares, 2000).

It was soon argued that the mammary gland was a logical choice for efforts to produce therapeutic proteins. First, milk proteins are expressed at high levels and with very good tissue specificity. Second, milk can be readily obtained and in the case of dairy species for long periods in high volume. The first report describing production of a recombinant human protein targeted the mammary gland for secretion of tissue plasminogen activator in mice. Use of mice has been critical to success because of the short generation time and the capacity to test the various gene constructs. Because of the expense and poorer success rate of producing transgenic livestock, rodent-based studies, despite the small quantities of milk produced, have allowed substantial testing of the biochemistry of secreted recombinant proteins so that better choices could be made for selection of constructs to be tested in larger species.

Success for these efforts, while evident, is far from routine. Much has been learned about the effectiveness of promoters for the major milk protein genes as tools to drive synthesis of recombinant proteins. Indeed, promoters from casein, β-lactoglobulin, or α-lactalbumin genes can all direct largely mammary-specific expression of a variety of transgenes. However, practically speaking, levels of production are variable even with the identical constructs. Another major problem is that the success rate is low for making transgenic animals with microinjection, especially with farm animals. The overall efficiency expressed in terms of the number of

Table 4.1 Expression of recombinant proteins in milk of livestock species

Species	Protein	Promoter	Expression
Sheep	hPC	Ovine βLG	0.3 mg/ml
	hFib	Ovine βLG	5.0 mg/ml
	hα₁AT	Ovine βLG	35.0 mg/ml
Goat	htPA	Caprine β-casein	6.0 mg/ml
	htPA	Human α-lactalbumin	2 mg/ml
Pig	hPC	Murine WAP	1 mg/ml
	mWAP	Murine WAP	2 mg/ml
Cow	hα-Lac	Human α-lactalbumin	2.4 mg/ml

Source: Adapted from Ayares, 2000, and Clark, 1998.
Note: Abbreviations are as follows: hPC = human protein C, hFib = human fibrinogen, $h\alpha_1AT$ = human alpha, antitrypsin, htPA = human tissue plasminogen activator, mWAP = murine whey acidic protein, hα-Lac = human α-lactalbumin.

eggs injected and transferred versus the number of transgenic animals produced is typically only about 2 percent. Irrespective of these limitations, to date at least 50 different transgenic proteins have been produced in milk with promoters from the ruminant α-lactalbumin, casein, β-lactoglobulin genes, or the murine whey acidic protein (WAP) gene. The proportion of these in livestock species is somewhat limited but growing. Table 4.1 illustrates some livestock species examples derived from transgenic animals produced with microinjection techniques.

Some workers have attempted to identify embryos that carry the integrated transgenes prior to implantation (i.e., to only transfer the positive embryos), but they have had only limited success. There is a flurry of recent work using cells in culture, rather than embryos, for integration of the transgene DNA. If the necessary genetic manipulation could be accomplished with cells in culture, potential donor cells could be isolated and returned to a host blastocyst so the cells and the blastocyst's descendents, especially germ cells, could become part of a fully developed offspring. This is exactly what has now been repeatedly accomplished with embryonic stem cells isolated from the developing embryos of several strains of mice. Unfortunately, similar systems for use in livestock species have not become available despite the efforts of many researchers.

However, scientists at the Roslin Institute in Scotland succeeded in producing viable cloned sheep from nuclei isolated from differentiated cells grown for many passages in culture. This unexpected discovery disrupted the long-standing developmental dogma that once a cell had differentiated, it could not be reprogrammed to function like a totipotent embryonic cell. However, these workers showed that either fetal or adult cells grown in culture could be reprogrammed to enter a resting state. This means that these cells, like embryonic stem cells, can be genetically altered in culture to ensure that the cell and its daughters contain the desired constructs and expression attributes before transfer. Like older cloning techniques, newly quiescent transgenic cells are used as nuclear donors for somatic cell nuclear transfers. With this method the donor cell is placed in between the zona pellucida and cytoplasmic membrane of an enucleated, unfertilized egg, and an electrical pulse is used to fuse the membranes of the cells. The reconstructed embryo contains the diploid DNA from the donor cell, and except for maternal contribution of mitochondrial DNA, the new animal is identical to the animal from which the donor cell is obtained. It is anticipated that these somatic cells can be manipulated in vitro to add or remove genes, similar to methods used with murine stem cells. This should allow the production of cloned, transgenic animals with well-characterized, predictable genetic modifications. Indeed, it was recently reported that three cloned transgenic sheep producing human blood–clotting factor IX have been made with these techniques. Clearly, there are many more advances to be made, but it is philosophically satisfying to consider that dairy animals long prized for making abundant, nutritious milk might also have a direct role in generation of pharmaceuticals for human health as well.

In addition to production of foreign proteins in milk, there is also interest in changing milk composition to better suit the milk-consuming public or milk product manufacturers. Milk composition can be altered by nutritional management (especially milk fat) or through exploitation of natural genetic variation. Historically, improvements in dairy cattle depended on crossbreeding to tap differences between breeds or selection to utilize variation within breeds. Genetic engineering in many ways is simply an extension of the art of these animal breeders. However, in dairy cattle the essential result of genetic selection has been virtually a continuous increase in milk production with relatively little change in milk composition. Answering the question of which genetic changes to select for altering milk composition will vary depending on global markets and consumer demands. At the present time in the United States, there is roughly a balance between supply and utilization of milk fat and protein, but lactose is in excess. In the British milk market, demand for fat production is higher, but protein is overproduced. These problems are confounded if consideration given to genetic manipulation better suits a specific sector of the dairy industry (e.g., cheese making) than the overall national economy. Clearly, there are concerns that go well beyond the development of laboratory techniques for genetic manipulation (Karatzas and Turner, 1997).

Until the recent breakthroughs with nuclear transfer and cloning of differentiated ruminant cells (see above), much of the discussion among scientists considered the possibilities for changing milk composition with the limitation of adding extra genes to the genome rather that considering the effects of gene knockouts or substitutions now theoretically possible. However, even the now relatively simple technique of gene copy additions could have dramatic impacts on milk composition and the functionality of milk. For example, cattle now have single copies of the casein genes in their cells. Because the caseins are the dominant milk proteins essential for cheese making, changes in the caseins could dramatically affect production of milk-derived foods. Cheese making involves aggregation of the casein micelles in a protein network that traps some water and much of the milk fat. As the cheese curd shrinks, whey (mostly water, lactose, and dissolved whey protein) is expelled as liquid. The final product is a mixture of caseins, fat, and water. For some cheeses (e.g., cheddar) the caseins account for half of the dry matter. Thus, the amount of casein in the starting milk is a critical element in determining the yield and nutritional attributes of the cheese. If the addition of an extra copy of the α-s_1-casein gene increased the content of α-s_1-casein in milk by 20 percent, it is estimated this would provide an increase of $190 million per year to the dairy industry because of a positive effect on cheese making. Similarly, increased β-casein in milk reduces rennet-clotting times and increases the extent of syneresis, again benefiting cheese making. In summary each 0.1 percent change in total milk protein (essentially caseins) increases milk cheese yield per 100 kg of milk by 0.16 kg. Expected greater cheese yields from the milk of Jersey cows (3.8 percent protein) compared with Holsteins (3.19 percent protein) illustrates the effect of even relatively small differences in milk composition (Karatzas and Turner, 1997).

With the development of techniques analogous to the use of embryonic stem cells for dairy animals, even more dramatic alterations in milk composition are possible. Essentially, these are ideas for (1) the modification of existing milk proteins to alter their function or nutritional properties or (2) removal of normal milk components to fit consumer demands. For example, among the caseins it might be possible to engineer a form of κ-casein more efficiently cleaved by the action of chymosin. Storage time needed for cheese ripening could be reduced by increasing cleavage sites in α-s_1-casein for the action of the enzyme chymosin. Although caseins are excellent nutritional proteins, increasing the content for methionine would increase their value since the caseins as a group are low in sulfur-containing amino acids.

Lactose intolerance is a problem for many milk and dairy product consumers. Indeed, reduced lactose milk and dairy products are available thanks to postharvest treatment of milk with lactases. A second route has the goal of reducing lactose by either the removal of the α-lactalbumin gene or by introduction of a lactase to cleave the lactose after secretion. The former has been achieved in transgenic mice. Mice with both alleles of the α-lactalbumin gene deleted produced very viscous milk with very high fat and protein content. However, pups could not survive because they were unable to effectively obtain milk from their mothers. These studies very dramatically confirmed the importance of lactose in regulation of milk volume via osmosis of water into the secretory vesicles and milk. However, modifications that would reduce but not eliminate lactose production might well be possible physiologically and could have a very large impact on the dairy industry. For example, simply reducing the water content by 50 percent would dramatically lower the volume of fluid milk needed to be stored on the farm and transported to processing plants. Savings in storage and transportation costs alone would be dramatic.

Although human breast milk is clearly the best choice for human neonates, many infants are fed formulas based on bovine milk. These formulas could be improved if the protein composition more closely resembled human milk. While seemingly only fantasy a few short years ago, some researchers have suggested it is feasible to knock out the bovine milk protein genes and replace them with human equivalents. In the shorter term it may be possible to make other modifications to improve milk used in infant formulas. For example, β-lactoglobulin, the most abundant of the whey proteins in bovine milk, is absent from human milk. Blocking the production of β-lactoglobulin in bovine milk would likely be a benefit since many of the allergy problems of human infants consuming bovine milk appear to involve β-lactoglobulin. Another example is human lactoferrin. Lactoferrin is a major whey protein in human milk, but it is normally very low in bovine milk. Bovine milk with added human lactoferrin could be beneficial. In fact transgenic cows producing human lactoferrin have been produced. Other modifications include increased secretion of lysozyme (a protein with antibacterial activity, normally found in human milk). Indeed, mice producing recombinant lysostatin (antibacterial protein) in their milk are protected in challenges with mastitis-causing microorganisms. Studies have led to the recent generation of transgenic cattle that produce the protein. If protection is successfully transferred to lactating dairy cattle, this technology could be a major breakthrough in mastitis control, which costs the dairy industry millions of dollars annually. Another example of a modification that would improve milk used in infant formulas is increased secretion of lipase, which would increase the capacity of infants to digest lipids. No doubt there will be successes and failures, but uses and attributes of bovine milk are surely destined to change in unexpected, even startling, ways.

Chapter 5
Milking Management

For a dairy enterprise, efforts to adequately feed, breed, and care for animals are wasted if milking procedures and milk handling are not satisfactory. After all, the essential product of a dairy is milk. This means that attention to methods of maximizing milk removal, preventing mastitis, and promoting general udder health is critically important. Cows are milked two or three times every day, and milking can be viewed as tedious, so diligence is essential to avoid developing careless milking techniques that can cause udder damage or disease, decrease yields and milk quality, and ultimately reduce profits. To minimize problems, careful attention to cow preparation for stimulation of milk ejection and the application and removal of teat cups is essential. This chapter provides an overview of the unique features of the udder (in contrast with mammary development generally), reviews important features of milking machines and milking techniques, and discusses the relative vulnerability of the udder to mastitis during milking.

Physiological Factors and Machine Milking

Preceding chapters provide an overview of mammary development, but the somewhat unique anatomy of the udder deserves special attention. In the cow and other ruminants, the mammary glands are clustered together into groups of two (goats or sheep) or four (cattle) to create the udder. This arrangement provides a practical advantage. Since the mammary glands and teats are close together, the portion of the milking machine attached to the animal (teat cups and teat cluster) can be relatively compact. For those not familiar with milking and management of modern dairy cows, the udder of a lactating Holstein cow, for example, can be rather massive. It is not unusual for a single cow to yield 25 kg or more of milk at a single milking. Combined with the mass of the udder tissues, this means that the connective tissues of the mammary glands have to support as much as 70 kg of tissue and stored milk just before milking. Given the ventral inguinal orientation of the udder, this is no trivial matter. Support is provided by strong, flat suspensory ligaments that are attached to the pelvic bone and to the strong tendons of the abdominal muscles in the pelvic area.

The udder is divided into two distinct halves, separated by the medial or median suspensory ligament, which provides most of the strength to hold the udder attached to the ventral body wall. Fibers of the lateral ligaments are continuous with the medial ligament but spread over either side of the udder so that it appears to the held in a sling of connective tissue. The medial ligament is somewhat elastic, but the lateral ligaments are not. As the milk accumulates in the udder, the normally vertical orientation of the teats is lost as teats progressively protrude laterally. As animals age, excessive degradation of the fibers of the medial suspensory ligament can reduce its support capacity so that the udder becomes pendulous irrespective of time relative to milking. This can lead to difficulty with milking (i.e., problems maintaining attachment of teat cups) as well as problems with teat injury and increased mastitis risk. Figure 5.1 illustrates the appearance and strength of the medial and lateral ligaments. The mammary glands of the udder are directly connected to the abdominal cavity only via passage through the inguinal canals. These are paired narrow oblique passages through the abdominal wall on either side of the midline, just above the udder. These canals allow passage of blood and lymph vessels and nerves to the udder.

Like other mammary glands, the udder contains two basic classes of tissues, (1) the parenchymal tissue responsible for milk synthesis and secretion and (2) the surrounding stromal tissue or connective tissue needed for support and to provide passage for blood and lymph vessels and nerves. Interestingly, although the two halves of the udder can easily be dissected by cutting along the medial suspensory ligament, there are no evident gross anatomical barriers between the front and rear glands (quarters) on either side of the udder. However, there are no direct connections between front and rear quarters. This is easily demonstrated following the injection of dye into the teat opening of one of the mammary glands. The dye stains only the tissue of the gland that is injected (Fig. 5.2). This means that the mammary glands of the udder are independent. This is sometimes an advantage in some experimental situations since one mammary gland or more often one udder half can be given an experimental treatment with the opposite side serving as a control. This of course is only relevant if treatments can be shown to have only local effects.

Teat Structure

Compared with most other mammary glands, those of animals with udders also have relatively large nipples or teats. Specifically, the teat of the ruminant has a single opening called the streak canal that leads directly into a space within the teat called the teat cistern. The structure of the streak canal is the primary defensive barrier against mastitis. The space of the teat cistern would typically hold only a few milliliters of milk. Near the base of the teat, there is an annular fold of tissue that separates the teat cistern from the gland cistern. The gland cistern is roughly the size of an orange and holds ~200 ml of milk. The gland cistern has

A

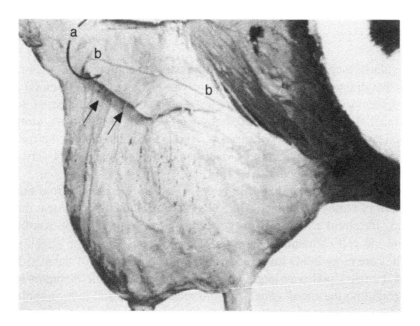

B

Fig. 5.1 Medial and lateral suspensory ligaments of the cow's udder. **A** shows the dissected udder supported by only the medial suspensory ligament (*arrow*). **B** shows a portion of the lateral suspensory ligament dissected as a flap of tissue (*arrows*). The letters *a* and *b* indicate the cut edge and boundaries of a flap of the lateral suspensory ligament. Adapted from Swett et al., 1942.

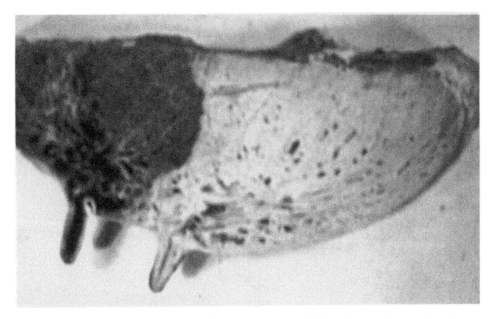

Fig. 5.2 Dye-injected udder. The right front quarter of this udder has been injected with dye. Notice that the color has remained only within the front quarter. Although there is no apparent demarcation between front and rear quarters, they are clearly independent.

many irregularly shaped cavities that accommodate the endings of large intralobular ducts that drain milk from the secretory tissue. Except for the terminal ducts directly adjacent to the alveoli, ducts are lined with at least two layers of nonsecretory epithelial cells. It is estimated that 40–60 percent of the milk is stored in the lumenal spaces of the ducts or cisterns. Above the gland cistern the tissue is progressively more dense and compact because of the relative lack of very large ducts and the closely arranged lobules (Fig. 5.2).

It has long been assumed that milk is prevented from escaping from the teat by the action of bands of sphincter-like smooth muscle cells in the teat meatus surrounding the streak canal. However, recent studies suggest that most of these smooth muscle cell elements are located some distance from the streak canal and that they are more likely involved in rhythmic contractions of the teat. Thus, closure of the teat canal may largely depend on passive elastic elements in the tissue surrounding the streak canal. A more recent suggestion is for a multispiraled, net-like combination of elastic fibers and associated smooth muscle cells that produce a spiraling of the internal epithelial folds of the streak canal to effect closure. Regardless of the exact mechanism, the internal structure of the streak canal and its surface secretions play a role in prevention of milk leakage as well as serving barrier functions. Although milking ease and milk speed are important factors in the economics of a dairy operation, selection of cows with wide, short teat canals may well speed the milking process at the expense of increased risk of mastitis.

The skin of the teat is hairless but tough and resistant to tears or punctures. Histologically, it is a stratified squamous epithelium that extends over the teat end and into the teat opening for the length of the streak canal. However, relative lack of insulation seems to make the teats susceptible to cold weather problems. The washing of udders and teats in preparation for milking and movement of cows into freezing temperatures and winds before drying likely amplify teat injury problems. Between the outside skin of the teat and internal surface of the streak canal or teat cistern, the stromal tissue contains a network of blood vessels, lymphatic vessels, smooth muscle cells, and nerves. The extensive blood supply is likely important in maintenance of normal temperature and especially so during severely cold temperatures. Mechanical milking can retard blood flow in the teat, produce vascular congestion, or induce swelling of the stromal tissue, which likely exacerbates problems with teat skin injury in cold weather.

Teat Opening, Teat Canal, and Teat End

The epithelial lining of the streak canal is also stratified squamous epithelium with a layer of keratin on the lumenal surface. The epithelium follows the contour of underlying dermal ridges, but as you traverse from the distal to the proximal end of the streak canal, the epithelium gives way to an area called Furstenberg's rosette. The convergence of the longitudinal folds or ridges of the teat canal serve to demarcate this area. The internal surface of the rosette is made up of 6 to 10 major folds of connective tissue covered by a stratified layer of nonsecretory epithelial cells two cells thick. Similarly, the epithelial lining of the teat and gland cisterns is characteristically stratified into a layer two cells thick. However, the subepithelial connective tissue throughout the teat contains a variety of leucocytes. Plasma cells and lymphocytes are the most common with the greater numbers in the rosette rather than in the teat cistern. Interestingly, the numbers of these cells are only slightly impacted by milking (i.e., in lactating vs. nonlactating cows). The streak canal ranges from 7 to 16 mm in length with a mean diameter of 0.82 mm (Fig. 5.3).

The lowest 2 cm of the streak canal is especially important because of the capacity of tissues in this region to act as a barrier to minimize milk leakage or entrance of environmental agents. It can be assumed that the diameter of the streak canal is positively related to the rate of milk flow. Cows with the best balance of acceptable rates of milk flow and protection from bacterial invasion can expect greatest longevity in the herd. Teats vary in shape and size, from cylindrical to funnel shaped. Teat ends also vary, with some that are flat and others round or pointed. Pointed teats are less common and are associated with slow milking times but resistance to mastitis. Round teats are common and occur on cows with faster milking times, but these cows exhibit some resistance to mastitis. Flat teat

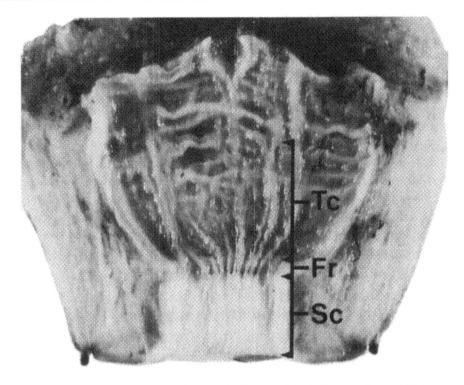

Fig. 5.3 Gross internal structure of the teat is illustrated, including streak canal (Sc), Furstenberg's rosette (Fr), and teat cistern (Tc). Adapted from Nickerson and Pankey, 1983.

ends are less common, but these cows also tend to be faster milking with less resistance to mastitis. Certainly, the uniformity of milking machine teat cups means that cows with teats of large diameter or length are more likely to suffer milking machine–related trauma.

Because the teat canal is lined with longitudinal folds, dilation of the streak canal during milking causes the epithelial lining to thin. This is somewhat analogous to the changes in transitional epithelium in the bladder. Regardless, this periodic stretch allows the keratin to spread over the surface to form a bactericidal barrier. With milking some of the keratin is flushed away during the periodic opening and closing of the teat canal, but fortunately it is constantly being renewed by the epithelium. The keratin itself has antibacterial agents that inhibit the growth of pathogens. Some researchers suggest that the minute areas of secretory tissue in the area of Furstenberg's rosette secrete protein(s) with bactericidal effects, but others suggest the material is lipid-like and made by the epithelial cells secreting the keratin. Certainly, the epithelial cells of the streak canal are constantly being renewed based on the appearance of mitotic cells in the basal layers of the epithelium (stratum germinativum). Passage of cannula through the teat canal or use of teat dilators scrapes away the keratin and can traumatize the epithelium. Experimentally, resistance to mastitis is markedly reduced if the keratin is removed. Studies in which pathogens were inoculated 3 mm into the streak canal caused infections in about one-third of treated glands. Inoculations 4 mm into the streak canal increased infection rates further, and inoculations 5 mm into the streak canal nearly always caused infection. Since pathogens, which cause mastitis, are not motile, to gain entrance into the parenchymal tissue, they must be moved by physical forces from the outside of the teat, through the streak canal, the teat and gland cistern, and the larger ducts, to the alveoli. Other than the period around milking, the keratin of the streak canal makes an effective barrier. However, animals with inherently thin keratin or animals with damaged areas are susceptible to local colonization with microbes and are therefore at greater risk from infection. During milking itself, retrograde movement of milk due to vacuum fluctuations or vacuum slips with leakage of air around the teat cups can allow bacteria-laden droplets to pass the streak canal.

Machine milking has a dramatic impact on the teat and teat end. Given the rate of milk flow of 7 to 8 ml/second, it is reasonable to expect that resulting shear forces might remove some of the protective keratin. It is also probable that some milk constituents are absorbed into the keratin during the time of milking or from milk droplets remaining after milking. If milking removes substantial amounts of the keratin and if renewal is delayed or changes in composition favor the formation of bacterial colonies or adherence, this could have marked effects on the streak canal as the primary defense against mastitis. In Holsteins, keratin weight before milking was 1.6 times greater than after milking (3.1 vs. 1.9 mg per teat). Jerseys by contrast were affected little by milking (3.5 vs. 3.1 mg per teat). Interestingly, there is a negative correlation between keratin loss at milking ($r = 0.53$, wet weight basis or $r = 0.65$, dry weight basis) and milk production. Total lipid in the keratin is similar before and after milking. In addition, although the major aspects of the fatty acid profiles are also similar before and after milking, keratin after milking has more short-chain fatty acids. This is consistent with the addition of milk-derived lipids to the keratin because of contamination of milk droplets remaining in the streak canal. In the case of the Holsteins, a greater proportion of the keratin is made up of these lipids after milking (Bitman et al., 1991). Although it was once thought that the keratin regenerated rather slowly (2–4 weeks), detailed quantitative studies show that following an initial collection, the keratin regenerates at a rate of 1.5 mg (wet weight) or 0.6 mg (dry weight) per day. This suggests complete restoration of the keratin occurs within 1 to 2.5 days (Capuco et al., 1990).

Historically, the appearance of teat ends was believed to be a good indication of mastitis health status. With the exception of damaged, traumatized teat ends with raw ulcerated regions, which do show a higher prevalence of mastitis, the appearance of the teat end seems to have little apparent relationship with infection status. Teat end appearance is, however, clearly affected by machine milking. After repeated milking the teat end tissue shows changes that initially appear as a ring around the teat opening. Hyperkeratosis of the teat canal is a physiological adaptation to action of milking. The progressive appearance of this callous ring around the teat canal opening is affected by days of lactation, teat length and shape, as well as parity. The teat end callus can be distinguished as a smooth or rough raised ring. After 1 week of milking, 55.7 percent of cows had no ring, but 41.4 and 2.9 percent of cows had a thin or moderately thick teat end callus. By week 13 of lactation, only 6.5 percent of cows had no visible teat end ring. Of those with rings 67 percent were smooth, and 33 percent were rough. Pointed or round teat ends showed more callous rings that were inverted. Longer time on the machine was also associated with a higher probability that the callous ring would be rough (Neijenhuis et al., 2000).

Milking Machine Components

Although it might seem obvious that the suckling calf or goat kid obtains milk by creating a vacuum around the teat, this in fact is not true. The calf compresses the neck of the teat between its tongue and hard palate and squeezes the milk trapped in the teat cistern into its mouth by compressing the teat between its tongue and the palate from the base of the teat toward the teat opening. The calf then lowers its jaws and tongue to allow the teat cistern to fill again, and the process is repeated. Consequently, sucking or creation of a vacuum within the mouth is not involved at all. This means that hand milking mimics normal nursing of the calf, since milk in the teat cistern is trapped by compression at the base of the teat and milk is pushed through the streak canal by sequential rolling of the fingers or stripping along the length of the teat. Regardless, calves, lambs, or kids fed with buckets or bottles with rubber nipples quickly learn to remove the milk by suction alone.

Although there is great diversity in the appearance of milking machines and milking systems, all have the same basic components. These include (1) a vacuum system, (2) pulsation elements, (3) an arrangement for transporting and collecting milk, and (4) a milking cluster that brings the vacuum and pulsation to the cow. A combination of milking machine components that allows more than one cow to be milked at a time is called a milking unit and is essentially a combination of the cluster and pulsation equipment. Related additional equipment is needed mainly based on how the milk from individual cows is combined for cooling and ultimately transport from the dairy. Three general arrangements are evident: (1) milking into a bucket or direct-to-can milking, (2) milking into recorder jars, which discharge into a pipeline, or (3) milking directly to a pipeline.

The basic milking unit for cows, the cluster, consists of four teat cups, each with a shell and a rubber liner or inflation; short milk and short pulse tubes; the claw; the long milk tube; and the long pulse tube (Fig. 5.4). Milk travels from the pulsation chamber to the claw piece, which receives commingled milk from each of the teat cups. Air is bled into the chamber (1) to displace the mass of milk that must be moved at any one time into a hose or pipe and (2) to speed the movement of milk away from the teat end into a larger vessel (milk bucket, weigh jar, or pipeline). Once the milk passes to the larger vessel, the milk and air fractions are separated. This allows milk, which is less compressible than air, to move mostly

Fig. 5.4 Milking cluster attached to a cow in a milking parlor. Teat cups, claw, and milk lines are evident.

by gravity to the milk receiver or storage tank. This separation also helps to minimize vacuum fluctuations that are associated with possible transfer of milk droplets between teat cups.

Within the teat cup, the liner is in contact with the teat and is continuous with the short milk tube, which empties into the claw. Rhythmic pressure acts to open the teat canal with periodic cycles of closure and relaxation corresponding with bouts of milk flow. These cycles are induced by variations in the vacuum applied between the outside of the liner and the wall of the teat cup shell. This is accomplished by the connections supplied by the short pulse tube. The pulsation rate is the number of times the air in the pulsation chamber of the teat cup is evacuated and returned to atmospheric pressure each minute. Pulsation rates of 40–160 cycles/minute have been evaluated in various milking studies. With machine milking the constant vacuum applied via the short milk tube at the teat end is essentially interrupted by the action of the pulsator. When the vacuum supplied to the outside of the liner is cycled off and atmospheric air enters, the stretch of the liner and the continuing vacuum at the teat end cause the liner to collapse around the teat. This closes the teat opening, compressing the teat wall, and partially occludes the liner below the teat. This temporarily stops milk flow, the teat is massaged, and blood that has accumulated in the tissue of the teat is stimulated to return to the udder. With the change in the pulsator, the cycle is reversed, and milk flow resumes.

A variety of liners are available, and makers make conflicting claims about what is needed in a well-functioning teat cup liner; however, there are certain desirable characteristics. Designs that reduce stripping time, teat cup fall-off, air slips, and machine "on time" are desirable. It is generally believed that liners that nearly close below the teat end during the rest or massage phase are also better. In many trials, good premilking stimulation improves yields and persistency, but there are apparent variations between breeds. Changes in pulsation ratio also impact milking and possibly long-term health of the teat. Common ratios are 50:50, 60:40, and 70:30. When the vacuum is applied for a longer period, the teat canal remains open so milk flow continues. This is likely to speed milking time simply because milk is being removed for a greater proportion of the milking cycle, but the amount of time spent massaging the teat is reduced. The weight of the teat cup cluster or, with some types of equipment, the increase of the down-ward pull by the teat cups minimizes yields of milk obtained by stripping near the end of milking and consequently reduces the need for stripping. Machine strip-ping of teats at the end of milking, the act of pulling downward on the milk clus-ter along with hand massage of the udder, is a widely used practice. This usually results in a brief increase in milk flow. Indeed, it is routinely considered a good milking practice to remove milk as completely as possible. The usual thinking is that incomplete milking reduces production and persistency. However, experi-mental data suggest that failure to machine strip actually has only a small effect on total production. Yields obtained with stripping vary but typically account for less that 2 percent of total yield. Certainly, the now common practice of milking three times per day means possible negative affects of retained milk would be even less. Whether machine or hand stripping at the end of milking, there is resid-ual milk that is not removed even with the most careful stripping techniques. This milk is only obtained by injecting the cow with oxytocin and milking again. The volume of residual milk is typically about 10–15 percent of the total. In practical terms, it is unlikely that machine stripping is a worthwhile practice. If addition of weight or inappropriate downward pull is applied to minimize the perceived need for stripping, this can also increase the chance that air will bleed in around the teat or that teat cups can come loose. Both of these contribute to increased incidence of mastitis. In short the milk obtained with stripping may be at the expense of increased risk of overmilking and mastitis (Heald, 1985).

Other problems can arise because of vacuum slips or teat cup slippage usually near the end of milking. When this happens, air rushes in around the teats, and the aerosol that can form acts to propel milk droplets and possibly bacterial-laden droplets through the cluster to either the milk hose or other teat ends when the vacuum is being maintained. Since the teat canal is open, it is possible that such droplets impact the teat end or even enter the teat cistern. If the teat is traumatized, introduction of contaminated milk is more likely to foster adherence of microor-ganisms and therefore formation of colonies of bacterial cells. This would likely increase the probability of generating new intramammary infections. This is more likely near the end of milking and if teat cups are removed before the vacuum is released. In fact, data showing that mastitis is reduced with milking machines equipped with automatic takeoffs may reflect the fact that the vacuum is always released prior to teat cup removal.

Machine Milking and Mastitis Control

Because of possible movement of milk droplets between quarters at milking or contamination of milking equipment between cows, milking procedures are an important part of management programs to minimize mastitis. As discussed in a subsequent section, mastitis is the most common and costly disease problem in the dairy. Not only are there the apparent losses in milk production when cows have acute mastitis episodes, but the often-unrecognized losses with subclinical mastitis are even more substantial. For example, it is estimated that it is 15 to 40 times more common than clinical mastitis. Moreover, it is of longer duration, is difficult to detect, and reduces milk production and milk quality. Clearly, cows are housed and maintained in environments with many sources of bacteria: the soil, bedding, and manure, for example, harbor many mastitis-causing organisms as well as microorganisms that can thrive in milk. Two immediate problems are apparent.

First, there is the need to minimize the contamination of milk. The most common indicator of milk quality is the bacterial count. Many states and cooperatives set limits of 10,000 or fewer organisms per milliliter of milk. The goal of the dairy industry is to supply high-quality milk to consumers. This job begins at the dairy. Filtration and pasteurization are important, but long shelf life and good taste are directly related to the quality of the milk that leaves the dairy. The care and hygiene necessary to prevent bacterial contamination of this milk are closely tied with management and hygiene practices to minimize cases of mastitis.

Proper milking begins with preparation of the udder for milking. Efforts to maintain cows in conditions that minimize amounts of soil or manure on udders or flanks at the time of milking should be encouraged. Regardless, udders should be cleaned and dried before the teat cups are attached. It is especially important that udders are dry. Water that runs down the flank and udder will tend to accumulate at the top of the teat cups. This accumulated contaminated water can easily be aspirated into the teat cup and either promote bacterial growth in the harvested milk or increase the chances of mastitis. Although the exact arrangement varies, a typical approach is to wash the lower portion of the udder with sanitizing solution to remove soil and debris and then to dry the udder with a single-service paper towel. Teats and teat ends should be thoroughly scrubbed with a paper towel, or a stream of water should be directed on them, and they should then be washed by hand. The udder should not be washed beyond 2 or 3 inches above the teats to facilitate drying. A more recent practice is to use a premilking teat dip in place of teat washing, if cows are not excessively dirty. Limited research data show this practice to be effective. To date, the only teat dips tested have contained iodophor, although other teat dips may prove effective. Regardless, the teat must be dried thoroughly before the milking unit is attached. Teats should not be predipped unless they are reasonably clean. A few streams of foremilk from each quarter are examined on the black surface of a strip cup. This allows the milker to detect early stages of clinical mastitis, removes foremilk that may have high bacteria counts, may serve as the primary stimulus for milk letdown, and assists in reducing mastitis. The appearance of flakes, clots, watery secretions, hard quarters, swelling, or redness provides an early warning of likely problems. As a general rule, there should be a sequence for udder preparation and attachment of the milk units so that milking begins about 1 minute after udder stimulation has begun. A regular routine should be established that results in smooth cow flow, uses good practices, and makes best use of milker time. Don't prepare too many cows in advance. Long stimulation times reduce fat tests, slow milking and increase milking time, and may increase somatic cell counts or mastitis problems. Maximum oxytocin concentration in the blood occurs shortly after the beginning of milking stimulation. Thereafter, oxytocin levels drop dramatically, and milk letdown is reduced.

Teat cups should be properly attached to minimize entrance of air into the system. Many research studies indicate that most of the infections associated with milking occur near the end of milking. This is linked to a greater chance for vacuum-related teat cup slips, as some teats milk out before others. The goal is to remove the machine just as the last quarter is milked out. If cows are properly prepared and milked with a good milking system, machine stripping should not be necessary. If practiced, machine stripping should take no more than 15 to 20 seconds. Each quarter should be massaged upward in a gentle motion with one hand while applying a slight downward pressure on the claw with the other hand. But it is important that there is not enough pressure to cause air to leak around the mouthpiece of the teat cup. The vacuum should always be shut off before the teat cups are removed. Improvements such as automatic detachers and robotic milking will continue, but hygiene and attention to detail are still essential at milking. The last step is to dip all of the teats in an approved teat dip.

After milking the teat cup liners may be contaminated with bacteria. Especially in herds where somatic cell counts average above 500,000, washing the teat cups between cows may be beneficial. For this the teat cups are usually rinsed in lukewarm water followed by a rinse in hot sanitizing solution. Teat cups are then hung inverted for several minutes so they can drain and dry. Teat cup liners must be dry before they are attached to the next cow.

Automatic backflush units are available and may be beneficial in certain dairy herds. As soon as the milking unit is removed from the cow, these systems automatically rinse the unit with water, followed by sanitizer and another rinse, before air is blown through them to dry liners. Backflush systems have been shown to reduce the number of bacteria present on teat cup liners and to reduce the contagious spread of mastitis pathogens via the milking system. However, backflush systems can increase the iodine content of milk because iodine can be retained in the polyvinyl chloride milk tubing and then released into milk.

The effectiveness of relatively simple milking hygiene measures to reduce mastitis incidence is not in dispute. For example, large British field trials in the late 1960s (Kingwill et al., 1979) were among the first to clearly demonstrate the effectiveness of hygiene at milking. These so-called full hygiene experiments included udder washing with boiled cloths or paper towels using 0.01 percent chlorhexidine, milkers wearing rubber gloves that were rinsed in disinfectant between cows, and milking machine clusters washed and dipped in sanitizing solution between cows. Compared with the usual control routine, the hygiene treatment reduced new mammary infections by 58 percent.

Monitoring Udder Health

Although strictly speaking, mastitis refers to any inflammation of the mammary tissue, the effects of microorganisms that invade the mammary gland cause most problems. Further, since inflammations are generally only detected among lactating animals, it is natural to assume that infections are mostly likely to occur in association with milking. In reality, cows are most prone to acquiring new intramammary infections (IMIs) in the period just after drying off and just before parturition. It is likely that such infections only become apparent after the onset of milking. Some scientific reasoning may explain this pattern of response. Soon after the mammary gland dries off, some of the mammary cells regress and die. Accumulated milk components also need to be reabsorbed. At least some of this process involves the actions of neutrophils and macrophages engulfing milk components and cellular debris. If during this period bacterial cells gain access to the mammary gland, these immune cells may be preoccupied with tissue remodeling so that microorganisms avoid destruction, thereby gaining an opportunity for colony formation. Second, immune responses generally are suppressed by treatment with steroid hormones. During the periparturient period concentrations of estrogens and glucocorticoids increase dramatically; this likely dampens immune responses. Further, although milking can expose the teat opening to possible exposure to microorganisms, if care is taken to avoid vacuum fluxes and good milking techniques are used, milk might well serve to flush microorganisms from the teat opening. Also, it is not uncommon for some animals to "bag up" during the final weeks prior to calving and even leak secretions. This suggests (1) that secretions accumulated on the teat end could provide nutrients to promote formation of bacterial colonies on the teat and (2) that the steak canal is at least temporarily open and susceptible to the entrance of microorganisms. In practical terms, it is also evident on most farms that dry cows do not get the close management and typical twice-daily observation of the milking herd. Last, neither do cows always give birth in the cleanest of environments. All of these factors likely interact to explain the increased prevalence of IMI in this period. This has led to the near universal use of intramammary infusions of long-acting antibiotics into each of the mammary glands at the time of drying off. This does not prevent all new IMI during the dry period, but this is an effective prophylactic procedure. Just as with antibiotic treatments during lactation, it is important to carefully follow recommended guidelines to avoid the possibility of antibiotic residues at the onset of milking postpartum.

Interestingly, it has also been found that care must be taken with the technique used in application of the product at drying off. Specifically, it is recommended that the teat ends be carefully cleaned and disinfected and that the treatment syringe cannula only be minimally inserted into the teat opening to infuse the product. Although either treatment technique (full vs. partial insertion of the syringe) effectively reduced appearance of new IMI in the dry period, new IMI occurred 2.5 times more often with the full insertion technique. The reason for this may be that the syringe cannula temporarily dilates the lumen of the streak canal, removes some protective keratin, or forces microorganisms from the streak canal into the teat cistern.

The somatic cell count of raw milk is the most common producer-related method to evaluate milk quality and udder health status of the lactating cow. Leucocytes and a small percentage of epithelial cells normally occur in milk. This combination of cells is referred to as the milk somatic cell count (MSCC). The term *somatic,* which means body, alludes to the fact that these are normal body-derived cells. By far ~98 percent of the cells are leucocytes, and most of these are neutrophils, sometimes called *polymorphonuclear leucocytes* (PMNs). This descriptive term is a reference to the lobed nucleus of these cells. Milk from uninfected cows typically contains less than 200,000 cells per milliliter, and it is not uncommon to find uninfected cows with MSCCs of 50,000 cells or less. Milk samples with values greater than 400,000 cells per milliliter are very likely from cows with inflammation probably caused by mastitis-producing organisms. These leucocytes enter the milk as a consequence of homing to the mammary gland from the bloodstream in response to chemicals released directly by bacterial cells or materials released by injured mammary cells. These chemicals induce chemotaxis that initially recruits neutrophils and thereafter macrophages (monocytes) into the udder. Since an increase in MSCC is closely correlated to an occurrence of IMI, the MSCC is measured in milk samples collected as a part of routine monitoring of milk composition in many dairy herds. However, it is important to remember that, strictly speaking, bacteria-induced mastitis can only be confirmed by the isolation of pathogenic organisms in aseptically collected milk samples by approved bacteriological methods.

Bacterial mastitis results when bacteria traverse the teat opening, overcome the defenses in milk, and begin to multiply. The likelihood for invasion is greatly increased if bacteria reside in or colonize the teat duct. Thus, routine use of teat dips after milking act to minimize this possibility. If microorganisms do successfully enter the udder, the ability of the bacteria to adhere to the wall of the teat or gland cistern improves the chances of infection. For example, two common mastitis-causing organisms, *Streptococcus agalactiae* and *Staphyococcus aureus,* readily adhere to the surfaces of the mammary ducts. Bacteria often initially act to affect local tissue areas in major ducts, and if the infection is not eliminated by the leucocytes, colonies form and, by multiplication and fluid movement, the organisms infect other areas of the mammary gland. The bacteria produce toxins that can cause swelling and ultimately death of ductular or secretory epithelial cells. Paradoxically, these events increase blood vessel permeability and the attraction of PMNs to the inflamed areas. These cells act to engulf the bacterial cells and destroy them, but the often massive migration of PMNs can disrupt the secretory epithelium, and dead and dying PMNs can also damage the epithelium.

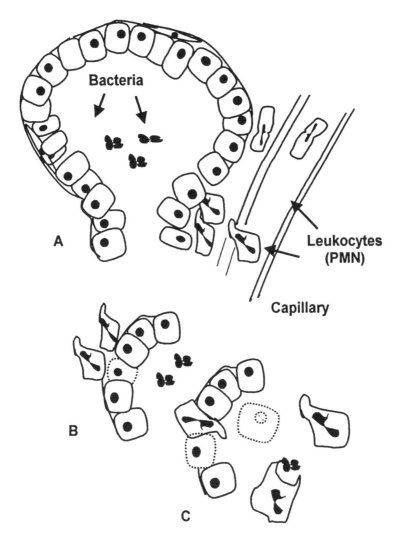

Fig. 5.5 Diagram of leukocyte response to presence of bacteria and toxins in the mammary gland. Chemotaxis causes PMNs to migrate from the capillaries and begin to pavement along the basement membrane of the infected alveoli (**A**). Some groups of epithelial cells become damaged by toxins or PMN activity and are sloughed into the lumenal spaces (damaged cells indicated by broken cell membranes). PMNs and macrophages accumulate in the alveolar spaces (**B**) where they engulf bacterial cells (**C**) and destroy them along with cell debris (Nickerson and Heald, 1981; Capuco et al., 1986).

After the PMNs cross the capillary wall, they move through the connective tissues toward the sites of bacterial invasion. They accumulate around the alveoli or ducts before entering the milk. The cells traverse the epithelium by either squeezing between the epithelial cells or by passing through areas where the epithelial cells have been killed. Figure 5.5 illustrates this invasion process. During the migration enzymes released by PMNs can also cause local destruction of secretory cells. Once the PMNs enter the milk, they depend on essentially random collisions to detect bacteria in milk for engulfment. This may explain the need for the massive migration of PMNs to sites of inflammation.

Unfortunately, it is generally believed that few damaged or destroyed secretory cells are replaced during a current lactation. Furthermore, accumulation of cell debris or clots or milk components (likely caused by acute local changes in pH or ion concentrations) can block drainage of milk from groups of alveoli and effectively induce local involution and loss of milk production capacity of these cells. Figure 5.6 illustrates an example of mastitis-induced alveolar cell damage, and Figure 5.7 the appearance of normal lactating tissue and adjacent area forced into involution by failure of milk removal.

Mastitis may occur in subclinical, acute, gangrenous, or chronic forms. Subclinical mastitis is the most prevalent, with 15 to 40 subclinical cases for every clinical event. During subclinical mastitis, the milk is usually visibly normal, but the concentration of MSCC is increased, or pathogens are identified from milk samples processed for bacteriological testing. The increased MSCC can be detected with a cow side test, such as the California Mastitis Test or Wisconsin Mastitis Test. Subclinical mastitis normally occurs before the other forms of mastitis and often leads to clinical mastitis. Acute mastitis is characterized by hot, painful, and swollen quarters or mammary glands and is frequently accompanied by fever and loss of appetite. On physical inspection the milk may contain flakes, clots, shreds, or blood. Milk yield is often severely depressed. In gangrenous mas-

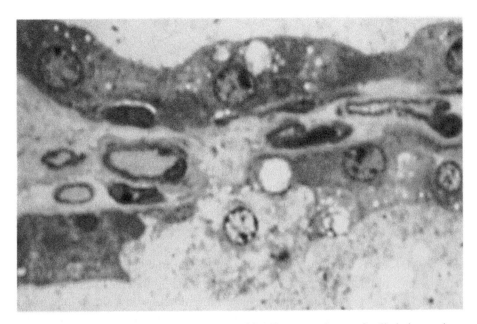

Fig. 5.6 Damaged epithelial cells after mastitis. The upper layer of cells is intact, but epithelial cells in the lower alveolus have been destroyed.

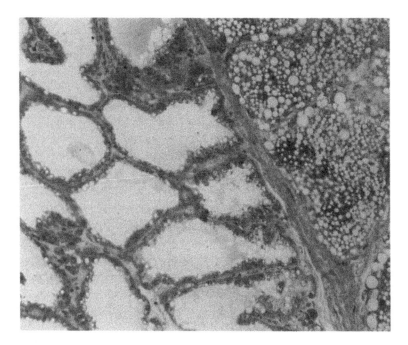

Fig. 5.7 Illustration of a normal, healthy lobule of mammary tissue (*left*) and an adjacent lobule (*right*) with milk stasis and involuting epithelial cells resulting from localized blockage of milk removal because of mastitis.

titis, the affected quarter may have a bluish discoloration and be noticeably cold to the touch. The discoloration most often proceeds from the teat upward. Chronic mastitis is characterized by repeated clinical attacks. The milk contains clots or flakes, and the quarter may be swollen. This can cause the glands to become hard and less pliable than normal. The chronically infected glands often fail to respond successfully to treatment, although the clinical symptoms may disappear temporarily.

Decreases in milk yield are associated with increasing MSCC. Such decreases are accompanied by changes in milk composition. In particular mastitis resulting from effects of major pathogens increases not only MSCC but also affects milk composition. Protein composition is markedly altered as the level of mammary-specific proteins declines and level of serum-derived proteins increases. The milk proteins also are more likely to be degraded because of the action of proteolytic enzymes. For example, the activity of plasmin can be more than doubled. Other enzymes released by damaged or activated PMNs also affect the properties of the casein micelles, and the proteolytic changes may continue even after the milk is removed. The content of casein, phosphorus, and potassium decreases while sodium, chloride, and free fatty acids increase. The latter is indicative of rancidity and can cause an unacceptable flavor. Not unexpectedly, the manufacturing properties of this milk are also altered. Table 5.1 summarizes milk constituent changes often observed in the milk of cows with mastitis.

Table 5.1 Changes in milk composition associated with mastitis and high MSCC

Component	Normal (%)	Mastitis (%)
Total protein	3.610	3.560
Total casein	2.800	2.300
Whey protein	0.800	1.300
Lactose	4.900	4.400
Fat	3.500	3.200
Immunoglobulins	0.100	0.600
Serum albumin	0.020	0.070
Lactoferrin	0.020	0.100
Sodium	0.057	0.105
Potassium	0.173	0.157
Calcium	0.120	0.040

Source. Adapted from National Mastitis Council, 1996.

Table 5.2 Relationship between MSCC, DHIA cell counts score, and milk production in dairy cows

MSCC	DHIA Score	Milk yield daily (kg)	Milk yield 305 days (kg)
12,500	0	29.2	8906
25,000	1	28.6	8723
50,000	2	28.0	8540
100,000	3	27.4	8357
200,000	4	26.9	8205
400,000	5	26.2	7991
800,000	6	25.4	7747
1,600,000	7	24.6	7503
3,200,000	8	23.6	7198
6,400,000	9	22.5	6863

Source. Adapted from Jones et al., 1984.

The National Dairy Herd Improvement Association (DHIA) has recommended that all DHIA computing centers report MSCC results using a scoring system, called *linear score,* which helps identify cows affected by subclinical mastitis and allows the producer to evaluate easily the effects of subclinical mastitis on milk production in the herd. The scoring system is based upon the relationships derived between milk yield and MSCC. Table 5.2 shows averages for cows within specific MSCC ranges for 33 herds sampled over 3 years in a study by Jones et al. (1984).

MSCCs below 50,000 per milliliter were common, and 37 percent of samples were from cows with less than 100,000 MSCCs per milliliter. There was a total of 57 percent of the cows with MSCCs below 200,000 and 77 percent below 400,000. Each lactating cow in these herds was also sampled for bacteriological testing every 3 months over 2 years. Only 5.9 percent of samples with MSCCs < 100,000 tested positive for the presence of pathogens. In cows with MSCCs ≥ 500,000 per milliliter, the percentage of cows infected increased to 25 percent. These data clearly show that cows with less than 100,000 MSCCs produced more milk and had fewer infections caused by major pathogens.

MSCC data are available to most dairy farmers through DHIA programs. When these results are combined with bacteriological culture results, factors most important for mastitis control on a specific farm can often be identified. As the data presented suggest, marked elevations in MSCCs are most frequently the result of subclinical or clinical mastitis caused by microorganisms. During inflammation episodes the majority of at least the initial increase in MSCCs is due to the rapid influx of PMNs. Milk from normal, uninfected glands generally contains less than 200,000 cells per milliliter. One recent study estimated that 50 percent of uninfected cows have MSCCs less than 100,000 and that 80 percent of these animals have MSCC values of less than 200,000. Certainly, the induced migration of PMNs into the mammary gland and milk can be rapid, and the number of cells in the first milking after response massive. Figure 5.8 illustrates changes in milk production and MSCCs in the milk of cows experimentally treated with an intramammary infusion of bacterial endotoxin. These results clearly show that the influx of cells does not depend on the presence of bacteria directly but rather that bacterial products can be potent chemotactic agents. In addition to the ~80 percent drop in milk yield by 24 hours after treatment, milk lactose concentration dropped from 5.1 to 2.3 percent and protein went from 3.3 to 4.9 percent. However, these effects were relatively short-lived since milk production and milk composition returned to normal by 72 hours after treatment. These results mirror many of the symptoms of an acute episode of clinical mastitis, especially the response to an environmental pathogen such as *Escherichia coli*. Figure 5.9 shows a histological section of mammary tissue of a lactating cow with an elevated MSCC. Numerous PMNs are present in the alveolar lumena.

MSCCs for milk samples from uninfected cows generally show only small changes with lactation number or as days within a particular lactation increase. For example MSCCs for samples from uninfected quarters averaged 83,000 on day 35 postpartum and 160,000 cells per milliliter on day 285 of lactation. In cows infected with *Staphylococcus aureus* the MSCCs went from 234,000 to 1,000,000 over the same period. Clearly, MSCC data are especially useful for monitoring herds experiencing problems with the contagious pathogens. This is because infections caused by these microorganisms tend to be of long duration, new IMI can lead to increased prevalence of infections among cows, and this may well be reflected in elevated MSCCs for bulk tank samples not just samples collected from individual cows. It is possible for well managed herds to have largely controlled mastitis caused by the

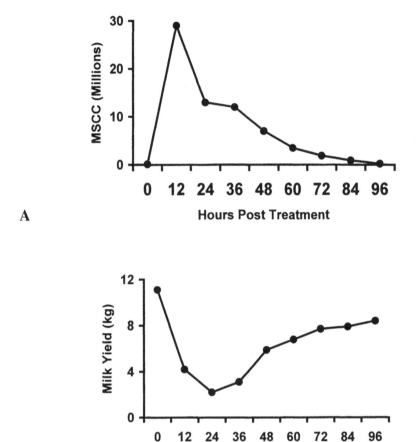

Fig. 5.8 Changes in MSCC (**A**) and milk production (**B**) in cows given an intramammary infusion of bacterial endotoxin. Adapted from McFadden et al., 1988.

contagious pathogens and to have problems with increased cases of clinical mastitis caused by environmental pathogens and yet maintain herd average MSCCs of 300,000 or less. This is because these infections tend to be relatively short-lived so that corresponding acute elevations of MSCCs in these cows can be missed with periodic sampling schemes and the effect on bulk tank MSCCs can be minimal. Such cases require careful sampling of suspect cows and attention to detail at milking to identify any early signs of clinical mastitis. It is frequently recommended that milk samples be collected and frozen for bacteriological testing from all quarters with clinical signs or from cows with elevated MSCCs.

Mastitis can be caused by a variety of microorganisms, but streptococci, staphylococci, or the coliforms cause ~95 percent of cases. Common mastitis pathogens found either on or in the udder are referred to as contagious since many of these do not survive in the environment. Environmental pathogens, in contrast,

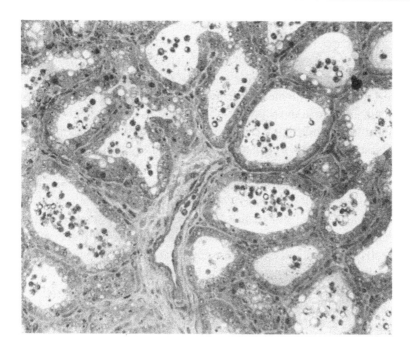

Fig. 5.9 Numerous leucocytes in the alveolar spaces of a lactating bovine mammary gland of a cow with a high milk somatic cell count (MSCC).

are routinely found in bedding, soil, and manure. Two prevalent contagious mastitis pathogens (e.g., *Staphylococcus aureus* and *Streptococcus agalactiae*) are spread from infected udders to noninfected udders during milking through contaminated teat cups, milkers' hands, towels used to dry the udder of more than one cow, or possibly by flies. While it is possible that new infections from environmental pathogens (*Streptococcus uberis, Streptococcus dysgalactiae,* or one of the coliforms such as *Escherichia coli* or *Klebsiella coli*) can occur during the milking process, it is more likely that the primary exposure to these pathogens is between milking times.

Regardless of the route for infection, identification of the specific pathogens causing mastitis requires that milk from the suspect gland be sampled by procedures that reduce the chance of contamination by bacteria from either the cow or the environment. The identification of the type of pathogen can be useful in determining the source of infection and consequently aid in developing procedures to minimize problems. Furthermore, antibiotic sensitivity testing on isolated colonies can assist the veterinarian in deciding the most appropriate treatment. Udders must be washed with a sanitizing solution and then dried because water droplets on teat ends may contain bacteria. Samples may be collected immediately after preparation for milking. The teat end must be disinfected with alcohol, and milk collected into sterilized tubes. Samples of milk are usually streaked onto blood agar culture plates and incubated at 37°C. Each of the major classes of microorganisms produces colonies of cells with characteristic features (e.g., size, shape, or general appearance). Thus, classes of microorganisms can be differentiated. Culture plates are usually read initially at 24 hours and again at 48 hours. Sometimes a second media is used along with blood agar to more fully differentiate certain species, or colonies are removed for further testing (i.e., the presence of a predicted enzyme activity or growth is tested in media supplemented with a specific nutritive substrate). It is also a routine practice to confirm the presence of a suspect mastitis-causing organism in replicate milk samples.

Streptococcus agalactiae most often infects mammary ducts in the lower regions of the udder but if unchecked can spread and cause tissue damage throughout the udder. Hyperplastic responses of duct epithelial cells can lead to thickening of duct walls, and if this occurs with the appearance of larger milk clots, it can clog smaller ducts. If these remain in place, milk accumulates, the secretory cells die, fibroblasts produce collagen and matrix proteins to replace the epithelial cells, and these areas of parenchymal tissue are no longer capable of producing milk. If the infection persists, *Streptococcus agalactiae* mastitis can become chronic, with periodic flare-ups and appearance of clinical signs. This infection usually is not associated with severe systemic disease, but repeated episodes can cause severe fibrosis of the infected mammary gland so that milk production potential is permanently impaired. *Streptococcus agalactiae* is frequently the cause of commonly occurring contagious, chronic mastitis. The source of infection is from other infected cows because the organism cannot live outside the udder. The use of common udder washing cloths or sponges is one factor associated with its transmission. Good hygiene, dry cow therapy, and teat dipping has eliminated this organism as a problem in many herds. Streptococci less frequently associated with mastitis include *Streptococcus uberis* and *Streptococcus dysgalactiae,* which are a reflection of environmental conditions. They do not depend directly upon the mammary gland for survival but can be found on udder skin, the mouth, or soil. Poor teat cleaning and excessive use of water with no or inadequate drying are frequent infection sources, as well as poor free-stall management, muddy lots, and poorly drained lots.

Staphylococcus aureus infections are often more damaging to secretory tissue because of the potent complement of toxins produced by these organisms. Furthermore, these organisms are prone to establish deep-seated areas of infection within the connective tissue of the alveoli. This can produce walled-off areas of abscess and, not surprisingly, induction of scar tissue. This walling-off phenomenon may explain the very poor antibiotic cure rates for *Staphylococcus aureus* mastitis. Often only relative small areas of secretory tissue are involved, but these cells frequently revert to a dedifferentiated state so that milk production is lost. Sometimes clogged ducts can reopen, but this can simply allow the shedding of microorganisms from walled-off areas to infect new areas of the gland. This can lead to the formation of multiple abscesses that can be palpated as lumps in the udder tissue. Occasionally, some strains of *Staphylococcus aureus* very rapidly multiply and produce α-toxins. These agents cause vasoconstriction and massive blood clotting so that oxygen and nutrient supplies to local tissue areas are completely lost. If this happens, the affected areas can express peracute mastitis with progression to gangrene. This can lead to complete loss of the mammary gland and, in extreme cases, death of the cow. Coldness of the affected tissue, a patchy blue discoloration, and leaking of blood exudate through the skin characterize gangrenous mastitis. Mastitis caused by *Staphylococcus aureus* is extremely difficult to control by antibiotic treatment alone. Successful control is often achieved only through prevention of infection and cow culling. *Staphylococcus aureus* organisms routinely colonize abnormal teat ends or can be present on teat cup liners. Milkers' hands and washcloths are ways in which the infection can be spread from cow to cow. The organisms probably penetrate the canal during milking. This strongly supports the often repeated recommendation that milkers should wear a new pair of disposable gloves at each milking and that gloved hands should be disinfected between cows.

Mastitis due to *Escherichia coli* is characterized by rapid onset and is more likely to be severe in animals or quarters with very low MSCC. The duration and clinical severity of the *Escherichia coli* infections seem to be associated with the degree of PMN response to the inflammation. For example, during very early lactation, chemotaxis of PMNs may be slower than in later lactation. This is believed to explain the severity of infections in early lactation. If the growth of the bacterial cells is not quickly controlled, the rapid multiplication of the cells leads to absorption of endotoxins into the blood and acute systemic responses (e.g., fever, changes in respiration, and appetite suppression). Milk becomes watery and yellowish with flakes and clots, and yield rapidly decreases. Fortunately, unless the strain is especially virulent, the influx of PMNs usually can destroy the microorganisms, and milk production returns to normal within a few days. The coliform group of organisms includes *Escherichia coli, Klebsiella pneumonia,* and *Aerobacter aerogenes.* Coliforms are considered environmental pathogens and are associated with factors such as dirty calving facilities, wet and dirty bedding, poor premilking practices, or even contamination during application of intramammary treatments. It is estimated that ~75 percent of coliform infections result in clinical mastitis (abnormal milk, udder swelling, and frequent systemic symptoms such as fever and loss of appetite). On the other hand, 60–70 percent of environmental pathogen infections exist for 30 days or less. Consequently, unless clinical signs are noted, the infections can be missed, particularly if milk samples for routine monitoring of MSCC are collected only monthly.

While it is appropriate that most attention has been concentrated on the major mastitis pathogens (*Staphylococcus aureus*, environmental streptococci, and coliforms), a variety of other bacterial and other microorganisms, so-called minor pathogens, can cause mastitis. For example a group of coagulase-negative staphylococci (CNStaph) can frequently be identified in milk samples. Even in well-managed herds with excellent mastitis control programs, these species may be the most prevalent microorganisms isolated and are estimated to infect ~15 percent of all glands. These include the following species *Staphylococcus chromogenes, Staphylococcus hyicus, Staphylococcus epidermidis, Staphylococcus hominis,* and *Staphylococcus xylosus.* Under these circumstances CNStaph microorganisms may cause 3–10 percent of clinical cases of mastitis. Nearly half of the milk samples containing CNStaph species had MSCCs less than 100,000, and only 10 percent of those isolated came from glands with MSCCs greater that 500,000. Fortunately, CNStaph infections are most often relatively mild.

Less frequently, a variety of other agents can also cause mastitis, including *Mycoplasma bovis, Pseudomonas aeruginosa, Serratia* spp., *Actinomyces pyogenes, Prototheca* spp., *Bacillus cereus, Nocardia* spp., as well as some yeasts and molds. Although rare, mycoplasma infections are contagious and can rapidly spread from cow to cow at milking. Milk from infected cows varies from watery to colostrum-like to tan with sand-like or flaky sediments. Response to antibiotic treatment is usually very poor. Other infection-associated problems can include increased abortions, low fertility, arthritis, and pneumonia. Unfortunately, specialized bacteriological techniques are needed to confirm these infections. Some of these agents are likely to gain entrance to the udder by use of contaminated intramammary treatment syringes (yeasts) or following injury (*Actinomyces pyogenes*).

In summary, mastitis costs the U.S. dairy industry ~2 billion dollars annually or about 11 percent of the value of the total milk produced in the United States each year. Costs attributed to reduced milk yields, milk discarded because of treatment or acute signs, and purchase of replacement animals are estimated at $102, $24, and $33 per cow per year. Direct costs (e.g., veterinary services, supplies, and extra labor) add an additional $13 per cow per year. Although mastitis can never be completely eliminated, costs can certainly be reduced. For example, for the average herd in Virginia, approximately $171 per cow per year is spent. This amounts to $18.6 million in costs to the Virginia dairy industry. A reasonable goal for well-managed herds would be to have an average MSCC of 200,000 and no more than three cases of clinical mastitis per 100 cows per month. If this was achieved for the average Virginia herd (128 cows), there would be a net increase in income of $57 per cow or an additional $7,296 annually. Certainly, the overriding principle in mastitis control is to reduce disease by decreasing exposure of teat ends to pathogens and/or increasing resistance of cows to infection. Regardless, attention to milk procedures, careful record keeping, and vigilance to decrease exposure to environmental pathogens remain critical aspects of any mastitis control program.

Chapter 6
Endocrine, Growth Factor, and Neural Regulation of Mammary Development

The endocrine system, perhaps more than any other physiological system, plays a central role in all aspects of mammary development (mammogenesis), onset of lactation (lactogenesis), and maintenance of milk secretion (galactopoiesis). Experiments beginning in the 1920s (Stricker and Grueter, 1928) showed that milk secretion could be induced in virgin rabbits by injecting a pituitary extract. In 1933 Riddle, Bates, and Dykshorn purified the protein responsible for the milk secretion response observed by Stricker and Grueter and named it prolactin (Prl). Today, the widely touted and utilized galactopoietic effect of somatotropin to increase milk production in lactating cows had its foundation in studies by Asimov and Krouze in the 1930s. They showed that injections of pituitary extracts consistently increased milk production in lactating cows. Scientists describing and quantifying the potent effects of the pregnancy on mammary growth and changes in the mammary gland at puberty spurred others to isolate and identify the steroid hormones estrogen and progesterone. Advances in purification techniques and understanding of steroid hormone chemistry allowed further studies leading to the production of these steroids for widespread animal testing.

Although the existence of mammogenic and lactogenic substances from the pituitary had long been known, the efforts of C.H. Li and colleagues in the 1940s to purify larger quantities of Prl and growth hormone (GH) were essential. Soon thereafter, specific roles for these hormones in the regulation of mammogenesis in rodents were delineated in classic ablation replacement experiments (Lyons et al., 1958; Nandi, 1958). In an extensive series of studies, triply operated (adrenalectomized, ovariectomized, and hypophysectomized) rats and mice were treated with various combinations of purified hormones to see if normal mammary development could be restored. Injections of estrogen and GH together caused proliferation of mammary ducts. However, treatments with estrogen, progesterone, Prl, and GH were needed for lobulo-alveolar development. The maximum ductular and lobulo-alveolar development, although still less than in pregnancy, was obtained in animals also given glucocorticoids. For some strains of mice, GH and Prl were both capable of stimulating lobulo-alveolar development. Interestingly, it was not until the 1960s that it was conclusively shown that human Prl and human GH were distinct proteins.

British researchers, focused on efforts to improve and maintain milk supplies during World War II, initiated many endocrine studies on mammary development and function in dairy animals (Cowie et al., 1980). For example, the effects of estrogen and progesterone on mammogenesis were extensively evaluated in attempts to induce lactation in nonpregnant animals. Although difficulties with needed surgeries and expense limit use of ablation replacement experiments to study mammary development and function in cattle, Cowie et al. (1966) studied hypophysectomized-ovariectomized goats and showed that mammary development comparable to that at midgestation could be obtained in animals treated with a combination of estrogen, progesterone, Prl, GH, and adrenocorticotropic hormone (ACTH). Such experiments served to confirm that at least the general effects attributed to these hormones on mammary development in rodents can be applied to mammary development in dairy animals.

Much of what has been learned since these pioneering studies is the result of advancements in technology. For endocrinology specifically, the development of radioimmunoassay (RIA) techniques in the late 1960s and early 1970s ushered in a golden age for the study of endocrine regulation of lactation and mammary development. Although bioassays had served to establish general themes (changes in pituitary Prl or GH and a general correspondence with major reproductive events, e.g., puberty, pregnancy, and lactation, or an increase in oxytocin at milking or suckling), widespread availability of RIA methods for Prl, GH, oxytocin, progesterone, and estrogen allowed study of hormone secretion on a scale previously unimagined. These techniques replaced the bioassays and allowed the accurate measurement of circulating blood or tissue concentrations of many hormones. Hormones and growth factors are often present in only picogram (pg) or nanogram (ng) quantities per milliliter of plasma. For the first time it became possible to determine correspondence between the secretion rate and pattern of secretion of a particular hormone and a specific physiological process. As an offshoot of methods for radiolabeling purified hormones for use in the RIA, isotope-tagged hormones were subsequently used to measure hormone and growth factor receptors on or in mammary cells. Despite the advancements allowed by use of the RIA, it is nonetheless important to remember that the method depends on antibody-antigen binding so that it is possible with highly specific antibodies (e.g., monoclonal antibodies) that fragments of hormones might be detected in addition to intact molecules. Since the method does not distinguish biologically active hormones, some caution in interpretation of results is also warranted. The potential problems led some researchers to develop radioreceptor assays (RRAs). In this variation of the technique, tissue fractions containing receptors for the hormone to be measured are used in place of the antiserum or antibodies in the RIA. The sensitivity of the RRA is often equivalent to that of the RIA, but it is rationalized that binding reactions and competition make a more physiologically meaningful measurement because specific receptors are used. Fortunately, in many cases correlations between results obtained with RIA and RRA are very good. The RRA technique is used less than the RIA because of the difficulty of producing and standardizing sources of receptors to be used in the assays. Others have used the growth response of the Nb2 lymphoma cells as a bioassay for Prl and other lactogenic hormones because the cells are very sensitive to addition of lactogenic hormones (Ellis et al., 1996).

It has become increasingly clear that, in addition to the steroid hormones, many of the peptide hormones also circulate in the blood bound to other proteins. For example, fractionation of serum by gel filtration chromatography and measurement of the fractions by RIA or RRA show the presence of GH or Prl activity in molecular weight fractions greater than the native hormones (~23,000 MW). Some of this likely represents the formation of hormone aggregates, but specific binding proteins for both have been identified. These are structurally identical with the extracellular domains of the receptors for the hormones. This suggests that these binding proteins are produced either by proteolytic cleavage of intact receptors from target cell surfaces or by alternative splicing of the mRNA transcript that encodes for the receptor. The role of these soluble binding proteins is unknown, but protection of the bound hormone from degradation or binding to cell surfaces would increase half-life in circulation. It is also possible that the hormone bound to these soluble receptor fragments could interact with other membrane-bound receptors to mediate presentation of Prl or GH to target cells. Regardless, since the RIA does not distinguish free from bound GH or Prl in the circulation, it is possible that simply measuring total hormone concentrations in circulation is a poor indication of the physiologically important fraction of the hormone. The situation for Prl is further complicated by the fact that Prl exists in several variants. For example, either as it is secreted or after interaction with target cells, some of the Prl is phosphorylated, glycosylated, or cleaved (Sinha, 1995). Some of these forms appear in circulation as molecular weight variants. Since the biological activities of these isoforms are not necessarily equivalent (Nicoll, 1997) like GH, measurement of total serum Prl probably tells only part of the story since a typical RIA, or RRA for that matter, is not likely to distinguish the isoforms. For example, it has been shown that the proportions of Prl occurring in the higher molecular fractions in serum may vary with stage of gestation. Specifically, in nonlactating Holstein heifers the proportion (~77 percent) of Prl that fractionates with higher molecular weight components is greater in early gestation (~100 days) than at 200 or 250 days of gestation. This suggests that the proportion of Prl available to the mammary cells during later gestation is also likely greater. This means that Prl would be more biologically effective during these periods even if the overall circulating concentration of Prl did not change. The technological evolution continues into the present with advancements in molecular biology and cell biology to decipher specific mechanisms for endocrine control of mammary cell function. It is also clear that despite the efforts of hundreds of researchers there is still much to learn about the role of Prl and GH in control of mammary growth and function.

Steroid Hormones and Mammogenesis

Because a majority of the studies on mammogenesis were completed before development of RIA techniques, there was no knowledge to determine if concentrations of the hormones resulting from treatments with exogenous hormone preparations were physiologically relevant or perhaps reflections of pharmacological effects of treatments. In retrospect, amounts of hormones injected likely often produced blood levels many times greater than during normal mammary development. However, it is equally clear that most of the broad conclusions were correct (e.g., the importance of GH and estrogen in control of mammary duct development). In their extensive monograph Cowie et al. (1980) summarized available blood hormone concentration data for estrogens, progesterone, glucocorticoids, Prl, and GH during periods of mammogenesis and lactation for classes of mammals ranging from the monotremes and marsupials to primates, rodents, and various domestic farm species. This section focuses on dairy animals and primarily the dairy cow. However, it is relevant to remember that there can be very marked differences between species. This is illustrated by comparing blood levels of estradiol in humans and cows. In human females estradiol averages about 50 pg/ml at the time of menstruation but peaks at about 300 pg/ml just before ovulation. In contrast in cows, concentrations average less than 2 pg/ml after ovulation with maximal concentrations of only about 15 pg/ml. In fact, it has been difficult to accurately measure circulating concentrations of estradiol-17β in the blood of virgin cattle, especially prior to onset of estrous cycles as well as at times during the estrous cycle. Concentrations of estrogens in artiodactyls seldom increase above 20 pg/ml except in late pregnancy. In primates concentrations range from 100 to 10,000 pg/ml, so in pregnant women concentrations of estradiol-17β can be 1000 times higher than for cows. Progesterone concentrations also tend to be lower in artiodactyls, usually less than 20 ng/ml. But in rodents, for example, concentrations are typically higher (~80 ng/ml), and in the guinea pig as high as 500 ng/ml. Progesterone concentrations in primates, like estrogen concentrations, are higher than for artiodactyls. This variation across species makes it difficult to relate hormone secretion patterns or specific circulating hormone concentrations with changes in mammogenesis. Furthermore, patterns of mammary development have been studied in detail in only a relatively few species.

In rats and mice, the shift toward allometric mammary development begins at roughly 3 weeks of age and is blocked by ovariectomy. In heifers, prepartum allometric mammary development begins at about 3 months and continues through the onset of puberty. Like the rodent, the ovary is essential for onset of prepubertal allometric mammary growth in cattle. Experiments by Purup et al. (1993a) confirmed the limited data from older studies and included detailed measurements of mammogenic hormones. Ovariectomy at 2.5 months of age markedly impaired mammary development and showed that without the ovary exogenous GH had no effect on mammary development. Specifically, GH treatment started at 176 days of age and continued for 15 weeks so that the animals were sacrificed at about 9 months of age just before puberty. Ovariectomy had only a small effect on serum concentrations of estradiol and GH. Average concentrations of serum estradiol (2.3. vs. 3.3 pg/ml) and GH (2.2 vs. 3.3 ng/ml) were both slightly lower in ovariectomized heifers before GH treatment. The pattern in older, placebo-treated animals was similar, but average concentrations of estradiol (3.4 vs. 4.1 pg/ml) and GH (5.3 vs. 5.5) were higher. Ovariectomy had no effect on circulating insulin-like growth factor I (IGF-I) either before or after onset of GH treatments. Neither did GH effect secretion of estradiol. However, mammary glands from ovariectomized heifers compared with those from intact heifers showed a reduction in total mammary gland weight and volume. Mammary parenchymal tissue mass determined by weight, volume, or chemical methods (DNA, dry fat-free tissue, or protein) was reduced in ovariectomized heifers. In intact heifers, GH significantly increased parenchymal tissue protein, and averages for all other measures of mammary parenchymal tissue were higher for GH-treated heifers.

Recent studies, reviewed in Akers et al. (2000), confirm the dramatic lack of mammary development in ovariectomized heifers and suggest that local mammary tissue changes in IGF-I axis molecules (IGF-I, IGF receptor, and IGF-I–binding proteins) are important for peripubertal mammary development and are altered by the presence of the ovary. Ongoing studies in which heifers were ovariectomized at approximately 1 month of age show that mammary parenchymal development is virtually arrested. There is no evidence of further parenchymal tissue development in ovariectomized heifers sacrificed at 6 months of age.

Regardless of the presence of the ovary, treatment of nonpregnant heifers with exogenous estradiol stimulates proliferation of the mammary epithelial cells. For example Figure 6.1 shows the proliferation index for thymidine incorporation into nuclei of mammary ductular cells after heifers were injected with estradiol-17β or progesterone or a combination of the two. Mammary tissue was collected by biopsy before treatments were administered and then after 24, 48, and 96 hours. Explants were prepared and incubated in culture media supplemented with tritiated thymidine. Explants were then fixed, embedded, sectioned, and processed for autoradiography. The percentage of epithelial cell nuclei incorporating the radiolabel was then determined as an index of cell proliferation. More that 85 percent of all labeled cells were epithelial. Estrogen increased cell proliferation 12-fold within 24 hours, with a further increase to 46-fold by 96 hours. Progesterone had no effect in the first 24 hours and relatively little effect (compared with estradiol) after 96 hours (6.2-fold). The combination of steroids was markedly less effective than estradiol alone. Capuco et al. (2000) have confirmed the potent effect of estradiol on mammary cell DNA synthesis in prepubertal heifers as noted by a marked increase in bromodeoxyuridine (BrdU) labeling of mammary cell nuclei in heifers injected with BrdU 2 hours prior to sacrifice. BrdU is an analogue of thymidine that can be detected in cells that incorporate the nucleotide using immunohistochemistry. Examples of labeled cells in mammary tissue from one of these heifers are depicted in Figure 6.2. Parenchymal tissue samples were embedded in paraffin or a plastic resin and processed to detect the BrdU. In this tech-

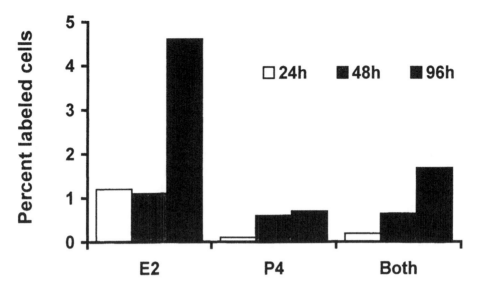

Fig. 6.1 Proliferation index of epithelial cells of mammary ducts for heifers treated with estradiol or progesterone or a combination of the two. A mammary biopsy of parenchymal tissue was taken before treatment and again after 24, 48, and 96 hours. Incorporation of tritiated thymidine was measured by incubating explants prepared from the biopsy tissue in culture media supplemented with tritiated thymidine. After 1 hour of incubation, explants were fixed, embedded, sectioned, and processed to measure autoradiographic localization of the isotope into epithelial cell nuclei (see Fig. 1.11A). Rate of incorporation prior to treatment averaged only 0.1 ± 0.05 percent. Estradiol clearly increased DNA synthesis. Adapted from Woodward et al., 1993.

nique labeled nuclei are remarkably similar to the appearance of autoradiograms for tritiated thymidine incorporation by mammary cells (Fig. 1.2).

The potent effect of estradiol on proliferation of the mammary epithelium when injected has been noted in many studies in a variety of species, yet precise mechanisms for the effect of estrogen, especially in cattle, remain unknown. In mice estrogen initially induces proliferation in the mammary stroma followed by proliferation of the mammary ducts. Yet in cattle neither adipocytes nor fibroblasts were stimulated by acute estrogen treatments (Woodward et al., 1993). This suggests that there may well be major differences in hormonal regulation of mammogenesis in rodents and cattle. Paradoxically, addition of estrogen to cultures of isolated bovine mammary epithelial cells or mammary organoids has no apparent effect on cell proliferation. This has led to the suggestion that effects of estrogen might actually be induced by the secretion of other mediators locally within the mammary gland or as a result of an estrogen-induced effect on another organ or tissue to secrete a growth factor (e.g., the pituitary gland). This is known as the estromedin hypothesis (Sirbasku, 1978). Furthermore, as shown in Figure 6.3 a modest but significant concentration-dependent increase in mammary epithelial cell proliferation

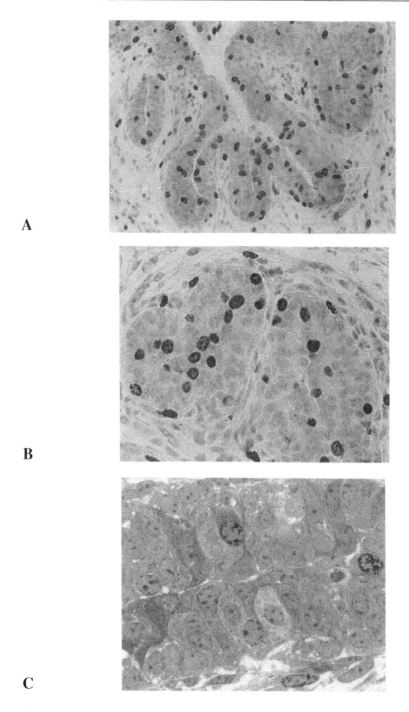

A

B

C

Fig. 6.2 Examples of bromodeoxyuridine (BrdU) -labeled mammary epithelial duct cell nuclei appear darkly stained as detected by immunocytochemistry for tissue from an estradiol-treated prepubertal heifer. **A** and **B** depict labeled cells for tissue embedded in paraffin with low (**A**) and moderate (**B**) magnification. **C** shows moderate magnification of similar tissue processed for embedding and processing in plastic. Unpublished images from Capuco et al., 2000.

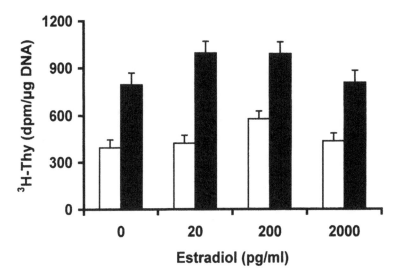

Fig. 6.3 Effect of estradiol alone (*open bars*) or estradiol + IGF-I (*closed bars*) on DNA synthesis in mammary tissue explants from prepubertal heifers. Estradiol alone has a small positive effect on DNA synthesis but improves the effect of IGF-I (300 ng/ml). Adapted from Purup et al., 1993b.

occurs in mammary explants incubated with estradiol (Purup et al., 1993b). The contrast with isolated bovine mammary epithelial cells may be explained by the fact that mammary parenchymal explants contain not only the epithelium of the mammary ducts but also stromal tissue proteins and cells associated with the ducts. Thus, mitogenic responses induced by estrogen may be mediated by autocrine and (or) paracrine secretion of growth factors, which then stimulate epithelial cell proliferation (Woodward et al., 1998).

On the other hand, the discovery of specific estrogen receptors via binding of radiolabeled estrogen in mammary tissue suggested that there are direct effects of estrogen in mammogenesis. Indeed, measurement of estrogen receptor concentration in human breast tumors remains a primary determinant for physicians to decide if a particular patient is likely to respond to hormone-specific therapies. Advances in techniques to measure estrogen receptors (or other steroid hormone receptors) have evolved, but presence of specific receptors is a requirement for a direct tissue effect for any hormone or growth factor. There is also increasing evidence for possible cross talk between growth factor and steroid hormone intracellular signaling pathways. Regardless, the data of Capuco et al. (2000) show the presence of numerous estrogen receptor–positive cells within the bovine mammary ducts of prepubertal heifers. Figure 6.4 depicts an example of estrogen receptor–positive cells in the mammary parenchymal of a prepubertal heifer. The majority of ductular epithelial cells express the estrogen receptor, and positive cells are found exclusively within the epithelium. Since even in response to injection with estradiol only a small fraction of the cells exhibit DNA synthesis, this suggests that expression of the estrogen receptor alone is not sufficient to induce proliferation even in the presence of exogenous estradiol. Further support for indirect mediation of the estrogen effect on mammary cell proliferation in heifers comes from a study of Berry et al. (2001) showing that heifers treated with estradiol have increased production of IGF-I in their mammary stromal tissue and that the stromal response is augmented in the stromal tissue from epithelial intact mammary glands. Last, it was recently confirmed that there are two distinct forms (α and β) of the estrogen receptor. The mammary gland appears to express primarily the α receptor, but it is possible that changes in expression patterns or ratios of expression of the variants are important in mammary development (Gustafsson, 1999). Changes in mammary tissue metabolism of estrogens are also likely important (Miettinen et al., 2000).

Despite the compelling evidence for the importance of the ovary for prepubertal mammary development of cattle, this is not a universal finding even for other ruminants. Ellis et al. (1998) reported the results of a factorial experiment to

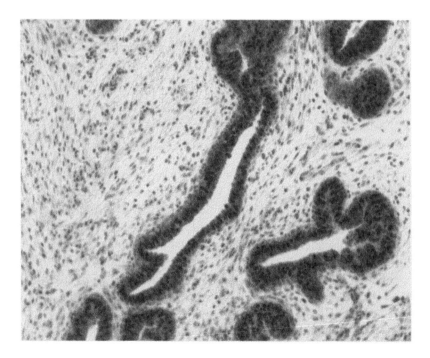

Fig. 6.4 Immunocytochemical localization of estrogen receptor in nuclei of mammary ductular epithelial cells of prepubertal Holstein heifer is shown. Note the abundant labeling of epithelial cell nuclei. Compared with the labeling to detect DNA synthesis (Fig. 6.3), there is little direct correspondence between presence of the estrogen receptor and induction of DNA synthesis following treatment with exogenous estradiol. Unpublished photomicrograph from study by Capuco et al. (2000).

assess the effects of ovariectomy, age, and estrogen administration on ovine mammary development. Lambs in the ovariectomy treatment groups had ovaries removed at 10 ± 1 day of age. After 6 and 13 weeks, groups of intact and ovariectomized lambs were sacrificed. During the week prior to sacrifice, lambs were administered estradiol-17β or a placebo. They also sacrificed a group of ewes at 12 days of age to provide reference data for initial mammary development. There were no significant effects of ovariectomy on total mammary gland weight or parenchymal or stromal tissue weights. Likewise, ovariectomy had no significant effect on concentrations of DNA, percentage of dry matter (DM), or percentage of fat in either parenchymal or stromal tissues. Indeed, all measures of mammary gland development were greater at 13 weeks than at 6 weeks and were similar in both ovariectomized and intact lambs (see Fig. 1.9). Clearly, in contrast with heifers onset of prepubertal allometric mammary development in ewes was not impaired by removal of the ovary. This interesting report is a potent reminder that results in one species do not necessarily translate to others and that regulation of development has been studied in only a small fraction of known mammals.

With the onset of puberty, the mammary gland in heifers continues to show an allometric rate of mammary growth for the first several estrous cycles but soon thereafter reverts to a period of isometric growth until conception. As Tucker (2000) reminded us in a recent review, "lest we become enamored with our current understanding of the hormones that control mammary growth and lactation, it remains a fact that the greatest physiological stimulus for milk yield is pregnancy, not some cocktail of exogenous hormones, growth factors, receptor agonists/antagonists, or gene therapies, Viva la mon!"

Steroids and Pregnancy

Although it should be clear that mammogenesis involves more than increased secretion of estrogen and progesterone, which occurs during pregnancy, this is nonetheless critical. For example, during the estrous cycle estrogen concentrations increase with follicular development and appearance of the dominant follicle, but with the subsequent ovulation concentrations decline and progesterone concentrations increase. Since estrogen and progesterone are both important for final duct growth and lobulo-alveolar development, the lack of a sustained simultaneous increase of both steroids during the estrous cycle probably explains the lack of marked parenchymal tissue development at this time. Furthermore, few of the studied species show evidence of lobulo-alveolar development prior to conception. With conception the corpus luteum is maintained and along with increasing production by the placenta in many species, blood concentrations of progesterone remain elevated throughout gestation. Concentrations of estradiol are higher (relative to the estrous cycle) with a further gradual increase during gestation and with a more dramatic increase during the final few weeks before calving. Consequently, concentrations of estrogen and progesterone are both simultaneously elevated during much of gestation, especially so during the later portion of gestation. This is believed to be responsible for much of the mammary growth during gestation. Prl and possible prolactin-like activity associated with secretion of placental lactogen is also important in mammogenesis in some ruminants (sheep and goats). In cattle, the role of Prl is most likely a permissive agent for the mammogenic effects of the steroids and other growth factors. For example there are no specific changes in secretion of Prl in cattle associated with mammogenesis and lobulo-alveolar formation during pregnancy.

An interesting accidental discovery reported by Smith and Schanbacher (1973) ushered in a flurry of activity on hormonal induction of lactation in cattle. Specifically, they were studying the effects of steroids on secretion of immunoglobulins in mammary secretions of nonlactating cows (i.e., animals that had failed to conceive). They observed that administration of estradiol-17β and progesterone for only 7 days, at doses that mimicked blood concentrations in animals near calving, caused the udders of some of the animals to "bag up." When these animals were milked, the initial colostrum-like secretions rapidly gave way to secretion of milk. As reviewed (Akers, 1985) subsequently, studies showed that lactation could be induced in about 70 percent of these nonpregnant cows and that milk yields for the successful animals averaged 70 percent of normal. Others noted positive correlations between the success of induced lactations and concentrations of Prl as well as improved yields when cows were also given drugs to induce Prl secretion. Also, greater milk yields measured for cows induced into lactation during the spring and summer were attributed to higher concentrations of Prl in serum compared with cows treated during winter months. Finally, goats or heifers simultaneously treated with ergocryptine (to block Prl secretion) failed to exhibit the usual mammary growth observed during induced lactation and thereafter produced relatively little milk. However, because Prl has a well-established effect to promote lactogenesis, it is difficult to distinguish effects on mammogenesis from those on lactogenesis in many induced lactation studies. In goats unilaterally mastectomized and hypophysectomized on day 60 of gestation, but given progesterone to maintain pregnancy, there was a five-fold increase in mammary parenchymal tissue on day 120 of gestation. In controls with intact pituitaries, mammary parenchymal tissue increased 10-fold. This indicates that the presence of the pituitary improves the growth response. Indeed, in other hemi-mastectomized, pituitary intact goats given ergocryptine, mammary growth was only increased five-fold. This degree of growth in the relative absence of Prl supports the idea that in goats placental lactogen is involved in mammogenesis during gestation. Changes in secretion of mammogenic hormones during gestation in cattle are illustrated in Figure 6.5.

Although changes in blood concentrations of mammogenic hormones are important in explaining changes in mammary growth, changes in tissue sensitivity and availability of biologically active hormones are also important. In circulation the steroid hormones are bound to transport proteins. Even in tissues the steroids can become sequestered with cellular lipids and thus become effectively unavailable (Capuco et al., 1982). This along with a decrease in progesterone receptor concentration and a change

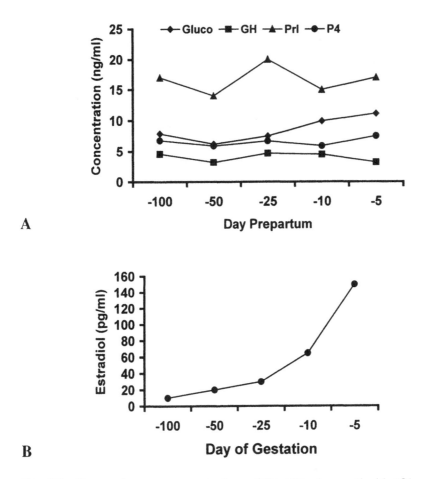

Fig. 6.5 Changes in serum concentrations of Prl, GH, glucocorticoids (Gluco), and progesterone (P4) in cattle are illustrated in **A**. Although concentrations of Prl vary substantially in response to photoperiod and temperature, stage of gestation has little effect. Once pregnancy is established serum concentrations of GH, glucocorticoids, and P4 are relatively stable during most of gestation. In contrast, concentrations of estradiol and related metabolites increase, especially in the later portions of gestation (**B**). Data adapted from Convey, 1974, and Tucker 1994.

in isoforms of the receptor is believed to explain the disappearance of the negative effect of progesterone on lactogenesis at the time of parturition in cattle. This illustrates the idea that the biological effectiveness of circulating hormones may change independent of the total blood concentration of the hormone. In addition, changes in expression, synthesis, or availability of hormone receptors in target cells impact biological response. While there is little data for expression of estrogen or progesterone receptors in bovine mammary tissue during mammogenesis, data from assay of ovine mammary tissue show that expression of the progesterone receptor occurs in close correspondence with appearance of lobulo-alveolar development (Table 6.1). Serum concentrations of

Table 6.1 Effect of stage of gestation upon mammogenic hormone concentrations and receptor concentrations in ovine mammary parenchymal tissue

	Day of gestation			
Measurement	*50*	*80*	*115*	*140*
Hormone binding	(n = 3)	(n = 4)	(n = 3)	(n = 4)
Progesterone (fmol/mg cytosolic protein)	125 ± 53	149 ± 26	656 ± 216	57 ± 22
Prolactin (fmol/mg microsomal protein)	7.2 ± 2.1	5.2 ± 1.9	32 ± 3.6	22.2 ± 2.9
Hormone concentration				
Progesterone (ng/ml)	3.6 ± 1.3	5.6 ± 1	29.9 ± 8.6	14.8 ± 0.7
Prolactin (ng/ml)	58 ± 10	24 ± 4	31 ± 8	134 ± 26
Growth hormone (ng/ml)	2.4 ± 0.4	4.4 ± 1.4	8.1 ± 1.3	15 ± 6

Source. Data adapted from Smith et al., 1987, 1989.

progesterone are consistently elevated during gestation with higher concentrations in later stages of gestation and in ewes with more than one fetus. As noted previously (Table 1.6), alveolar formation has begun by day 115 of gestation in sheep.

Anterior Pituitary Hormones and Mammogenesis

As its name might suggest, Prl has undoubtedly been the most intensely studied hormone related to lactation and mammary growth. Despite the understanding that the presence of the pituitary is essential for normal mammary development, whether Prl or GH predominates in mammogenesis is not clear. The answer to this question is likely species-dependent. For rodents stimulation by Prl is needed for both mammary growth and functional differentiation of the alveolar cells during pregnancy. Prl induces regression of the fully elongated terminal end buds and promotes appearance of ductular side branches via indirect systemic effects in the virgin mouse (Das and Vonderhaar, 1997). However, increased lobulo-alveolar development during pregnancy is a direct effect of prolactin binding in the mammary gland (Henninghausen et al., 1997). Experiments that disrupt the normal Prl-signaling cascade so that one of the family of signal transducers and activators of transcription (STAT5a) is inactivated result in lactation failure (Hynes et al., 1997). This is directly related to impaired development of the terminal end buds. Interestingly, since cows lack typical terminal end buds (see "Tissue, Cells, and Organization" in Chap. 1), it may be that Prl is relatively less important in mammogenesis in cattle, especially prior to lobulo-alveolar formation. It is well-known that Prl secretion is markedly inhibited in cold weather and during short photoperiods, yet mammary development is essentially identical for calves born throughout the year. At a minimum this suggests that wide variations in circulating concentrations of Prl have little impact on mammogenesis.

In the past 20 years, it has been established that native Prl is really part of a family of structurally related protein isoforms or variants. The Prl gene is transcribed not only in the lactotropes of the anterior pituitary gland but also by cells in the placenta, hypothalamus, and mammary gland and by lymphocytes. Moreover, either as it is secreted or after interaction with target cells, some Prl is enzymatically modified to become cleaved, phosphorylated, or glycosylated. Thus, the pleiotropic actions of Prl may ultimately be attributed to the presence of these different isoforms of the hormone. As suggested by Das and Vonderhaar (1997), the signaling pathway for Prl stimulation of cell differentiation compared with that for Prl signaling of cell proliferation needs to be deciphered to truly appreciate the role of Prl in mammary cell growth compared with differentiation. Possibilities include (1) differing effects of the Prl variants, (2) differences in Prl receptor subtypes, or (3) that the signaling pathways for these responses are distinct.

At least three distinct forms of the Prl receptor are also known to exist (Goffin and Kelly, 1997). These single-chain membrane-spanning proteins are arranged into three domains: (1) an extracellular region where Prl binding occurs on the extracellular surface, (2) a hydrophobic portion that spans the plasma membrane, and (3) an intracellular cytoplasmic domain. The three forms of the receptor exhibit differences in their cytoplasmic domains. The long form is 90 kDa and differs from the short form (40 kDa) because of differential splicing of mRNA transcribed from the Prl receptor gene. An intermediate version of the receptor is a deletion mutant of the long form that lacks 198 amino acids in the cytoplasmic domain. The intermediate form of the receptor is more sensitive to Prl and is a major form of the receptor found in the Nb2 rat lymphoma cells that are used as bioassay for lactogenic hormone activity. This may explain why these cells are so sensitive to addition of Prl and thus the utility of the cells as a bioassay tool. Prl also belongs to a superfamily of structurally related proteins that include GH and placental lactogen. The receptors for Prl and GH are single-chain proteins with one transmembrane domain (see Fig. 2.6). Structural features of the receptors indicate that they are part of what are currently called class 1 cytokine receptors that include receptors for several interleukins, erythropoietin, leptin, and others. This realization has facilitated the study of intracellular-signaling pathways since the structural similarities suggest parallel similarities in mechanisms of action. Expression of the Prl receptor in the bovine or ovine mammary gland increases dramatically near the time of parturition in concert with lactogenesis, and this level of expression is generally maintained during lactation. But there is no evidence for expression of different forms of the receptor to suggest that possible mammogenic versus lactogenic effects on mammary growth or function in ruminants are associated with a particular Prl receptor subclass (Smith et al., 1993).

Other Hormones and Mammogenesis

Although the ovarian steroids along with GH and Prl are certainly the major stimulators of mammary growth, several other of the classic hormones play permissive or secondary roles in mammary development. In addition to the ovary and corpus luteum, the placenta produces estrogen, progesterone, and a prolactin-like hormone placental lactogen (PL). As the name suggests, PL was first recognized by its Prl-like biological effects. However, the PLs of many species have both GH and Prl-like activities. For example, Prl and nonprimate GH molecules interact only with their specific receptors, called *lactogenic* and *somatogenic*, respectively. Human GH, in contrast, recognizes both receptor classes. Coincidently, this probably explains the inappropriate breast development sometimes observed in human males with pituitary tumors that overproduce GH. Both ovine and bovine PLs behave like human GH—they compete for both somatogenic and lactogenic receptors. This also explains some of the names associated with these proteins isolated from the placenta (e.g., chorionic somatomammotropin or chorionic mammotropin). Although PL has been implicated in preparation for the mammary gland for lactation, stimulation of steroidogenesis, fetal growth, and alteration of maternal metabolism, direct evidence for effects in cattle is lacking. The best evidence for the role of PL in mammogenesis is in rodents and in sheep and goats. Rats and mice have been shown to produce two forms of PL (PL-1 and PL-2). PL-1 appears early in gestation, peaks, and remains stable for most of gestation. PL-2 is produced initially at about midgestation but then peaks near parturition, supporting the idea that it is involved in lactogenesis. Evidence for the presence of a PL in goats and cows first arose from studies in which pieces of placental cotyledons were cocultured with mouse mammary tissue explants and this induced lactogenesis in the mammary tissue. Ovine and bovine PLs were subsequently isolated, and assays developed. In goats and sheep concentrations of serum PL are also higher in dams carrying twins and correlate with mammary development and subsequent milk production. In the bovine most of the PL is secreted into the fetal circulation so that concentrations in maternal serum are markedly lower than for sheep or goats. If the major maternal effects of PL are assumed to be lactogenic, it is unlikely that the low concentrations in serum of the pregnant heifers could effectively counteract the antilactogenic actions of high concentrations of progesterone at this time. Furthermore, exogenous recombinant bovine PL given to lactating pregnant or lactating nonpregnant cows had little effect on maternal metabolism (Byatt et al., 1992). Although consistent with a GH-like activity, concentrations of IGF-I were increased. Exogenous PL has a positive impact on mammary development and milk yield in heifers induced into lactation (Byatt et al., 1997). However, it is likely that naturally occurring PL in the pregnant cow is more important in fetal development than in mammary gland development or differentiation.

Adrenal steroids are essentially for physiological homeostasis, but it unlikely that these hormones are directly critical for mammogenesis. Thus, their role is secondary to the pituitary and ovarian hormones. Injections of glucocorticoids can stimulate development in immature animals but adrenalectomy does not consistently reduce mammary development. Similarly, the thyroid gland is important metabolically, but thyroidectomized animals can conceive and lactate, so this means thyroid hormones are not essential. However, thyroid hormones enhance the effectiveness of the ovarian steroids and thereby mammary growth during gestation. Mammary duct development is also impaired with hypothyroidism. This suggests that the thyroid hormones, like the adrenal steroids, allow for optimal mammary development and function. Relaxin is a protein hormone structurally related to insulin and IGF-I and produced in the corpus luteum. As the name indicates, its traditional function was associated with relaxing (via induced secretion of local proteases) of the pelvic ligaments to minimize birth problems. There is some evidence that it can enhance mammogenesis in rats and pigs, but work with the protein has been minimal because of the lack of purified hormone, especially related to cattle. Parathyroid hormone is associated with calcium homeostasis and therefore indirectly related to mammogenesis. On the other hand a protein produced by the alveolar cells of lactating animals, parathyroid hormone–related protein (PTHrP), seems to play a significant role in stimulation of mammary uptake of calcium. This is certainly important considering the very large calcium demand associated with milk production.

Local Tissue Mediators of Endocrine Action in Mammogenesis

An integral, often ignored, constituent of the mammary gland is the stromal tissue. For example, the development of the epithelium in the mammary gland of the fetus is influenced greatly by the surrounding mesenchymal tissue. In fact, androgen effects on the stromal tissue mediate the sexual dimorphism of mammary development between males and females. In the postnatal female mammary gland, the stroma exists as a matrix of connective and adipose tissue, collectively termed *mammary fat pad*. The mammary fat pad and its various constituents facilitate many of the gland's functions: it houses a vascular and lymphatic system, it provides a three-dimensional matrix and basement membrane, and it functions as a local site of hormone action, lipid biosynthesis, and growth factor synthesis (Hovey et al., 1999).

Studies in the past 10 years have begun to unravel roles for both systemic and mammary tissue–specific growth factors believed to mediate many of the effects of the "traditional" hormones. Among these are transforming growth factors (TGF-α and TGF-β), insulin-like growth factors (IGF-I and IGF-II), epidermal growth factor (EGF), and fibroblast growth factor (FGF). Other unrecognized growth factors are also likely involved in mammary development. Since mammary growth is tightly controlled, it should not be surprising to ultimately learn that regulation likely depends on a fine balance between growth stimulators (the common sense definition of a growth factor) as well as growth inhibitors. In many respects TGF-β(s) appears (appear) to serve such a negative function. For example, implants containing TGF-β1, -2, or -3 act locally and reversibly to inhibit ductal growth when placed near actively growing end buds in mice (Daniel et al., 1996). Moreover, peripubertal expression of a TGF-β1 transgene in mice prevents subsequent lobulo-alveolar formation during pregnancy and consequently prevents lactation (Smith, 1996).

Mechanisms are not clear, but inducing early senescence of duct cells may act to reduce a population of epithelial cells destined to be the direct precursors of the secretory cells of the alveoli. These data lead to the hypothesis that proliferation of ductal cells in the prepubertal mammary gland gives rise to two functionally distinct groups of progenitor cells: one capable of producing daughters committed to ductal formation and another committed to lobulo-alveolar formation and milk secretion (Chepko and Smith, 1999). How this hypothesis applies to ruminants is unknown, but TGF-β1 receptors are expressed in mammary tissue from peripubertal heifers and pregnant and lactating cows (Plaut, 1993; Plaut and Maple, 1995). The following discussion will focus on several growth factors shown to have an impact on mammary development in cattle and sheep. Much of the work related to ruminant mammary gland growth is at its beginning stages. A common model for these efforts has been to study the effect of purified growth factors on the growth of mammary epithelial cell organoids isolated from the parenchymal tissue of prepubertal heifers. In these experiments growth of cell organoids or clumps is measured by determination of the DNA content or changes in DNA synthesis for cells cultured in three-dimensional collagen gels. This culture system more closely mimics the "natural" growth of mammary cells than, for example, monolayer cell cultures because the cells of the former can develop outgrowths or extensions of cells that are morphologically reminiscent of mammary ducts. However, it is still important to appreciate that responses may not necessarily mirror growth factor effects in the intact mammary gland. It is critical that animal experiments continue to be used to validate ideas and hypotheses derived from culture studies. Nonetheless, in vitro studies are valuable research tools to decipher mechanisms for growth factor mediation of mammary development.

The Insulin-like Growth Factors Axis

As in other species, the prepubertal bovine mammary gland grows in response to exogenous treatments with either GH or estrogen. While effects of complete removal of endogenous GH on mammary development in the bovine have not been reported, a positive effect of GH on udder growth in heifers has been confirmed in several studies with cattle and sheep. However, GH-induced increased growth before puberty does not necessarily translate into increased milk yield. In contrast, limited studies suggest that GH treatment during late pregnancy stimulates both mammary growth and increased milk yield during the subsequent lactation. The limited data for effects of GH on mammary growth during lactation indicate that mammary growth is unaffected by GH treatment in early lactation, whereas GH seems to increase mammary parenchyma mass in midlactation (Sejrsen et al., 1999).

Specific receptors are needed for GH to have a direct effect on the mammary gland. However, it has been difficult to demonstrate the presence of specific GH receptors in pubertal ruminant mammary tissue using ligand-binding assays. The same is true with mammary tissue from pregnant or lactating cows or sheep. On the other hand, the ruminant mammary gland does express mRNA for the GH receptor (Hauser et al., 1990; Glimm et al., 1992). Moreover, Sinowatz et al. (2000) have shown extensive expression of both the mRNA by in situ hybridization and of the receptor protein by immunocytochemistry. These data, in contrast to the ligand-binding assays, do support the idea that the GH receptor has a direct role in mediation of GH effects on the mammary gland. Regardless, a widely held hypothesis is that at least part of the effect of GH on mammary growth is mediated by IGF-I. Thus, the mechanism of action of GH remains a puzzle but almost certainly involves the IGF family of related proteins, as well as locally produced factors, including receptors, binding proteins, and perhaps other growth factors (Akers et al., 2000).

IGF-I and IGF-II are widely expressed endocrine-, autocrine-, or paracrine-acting peptides that regulate cell growth, cell differentiation, maintenance of cell function, and prevention of apoptosis in multiple cell types. Research with IGF-I and IGF-II was initially centered on the somatomedin hypothesis (Le Roith et al., 2001), which proposed that these growth factors were the endocrine mediators of somatotropin (GH) effects on postnatal growth. Since these early experiments, with the somatomedins now called the IGFs, the view has evolved that the IGFs are important local-acting autocrine or paracrine stimulators of cell function. As with other growth factors, these peptides can interact with several related cell surface receptors (Adams et al., 2000; Butler et al., 1998). The primary signaling receptor for IGF-I (IGF-IR) is a tyrosine kinase receptor structurally similar to the insulin receptor. Members of this family share a heterotetrameric $\alpha_2\beta_2$ structure. The α and β domains and a signal peptide are synthesized as a 1367-amino acid long precursor transcribed from a single mRNA species. The $\alpha\beta$ heterodimers are produced by proteolysis of the precursor protein and linked by disulfide bonds. Two of these units are then linked by secondary disulfide bonds to create the mature $\alpha_2\beta_2$ receptor. IGF-IR binds with IGF-I with high affinity (Kd ~1 nM) in bovine mammary tissue, but affinity for insulin binding is about 500 times lower. In the cow, normal circulating concentrations of insulin 1–5 ng/ml would have little ability to signal via IGF-IR. This is relevant because most classic culture experiments with explanted mammary tissue utilized additions of 500 to 1000 ng/ml of insulin as part of the media to maintain the tissues in culture. Consequently, effects originally attributed to insulin were likely caused by insulin binding to IGF-IR. IGF-I can also bind to the insulin receptor but with much less affinity (~1000-fold). The situation is further confounded by the existence of hybrid receptors between IGF-IR and the native insulin receptor (IR) that have higher affinity for IGFs than for insulin. IGF-II is known for binding with high affinity to a receptor that is identical with a receptor for mannose-6-phosphate, but the receptor has no known intracellular signaling function. The affinity of this receptor for IGF-I is about 100-fold lower than for IGF-II, and it does not recognize insulin. Interestingly, there is evidence that IGF-II, but not IGF-I, can stimulate growth of cells by binding to a splice variant of the IR called IR-A. There are as many as five distinct types of insulin IGF-I hybrid receptors. IGF-IR and the IR isoforms IR-A and IR-B have the ability to form both homo- and heterotetramers. This can markedly change the diversity of signaling induced by binding of insulin or IGFs. Signaling pathways and biological effects are best characterized for IGF-IR and IR-B, but experiments in cell culture model systems support the idea that combinations of native and hybrid receptors allow both overlapping and unique physiological effects. In general stimulation of either IGF-IR or IR-B is associated with cell cycle progression, but stimulation of IR-B is more closely related to metabolic events, and stimulation of IGF-IR with mitogenesis. These major properties were illustrated in elegant molecular studies in which the cytoplasmic domains for IR-B and IGF-IR receptors were swapped. In the normal situation, there are overlapping as well as specific signaling events associated with activation of IR-B or IGF-IR (Hadsell and Bonnette, 2000).

For IGF-I binding to its receptor to be effective, two cellular processes must be meshed together. First, the binding reaction must transmit a signal through the plasma membrane to regulatory molecules located on the cytoplasmic face of the membrane. Second, a signal is needed to cause localization and interaction of the internal receptor domain with downstream effector molecules of the signal transduction cascade. Briefly, upon ligand binding, the IGF-IR cluster and intracellular tyrosine kinase is activated. This produces autophosphorylation and transphosphorylation of the β subunits of the receptor. When phosphorylation of certain tyrosine or serine amino acids of the cytoplasmic subunit occurs, this creates docking or binding sites for intracellular signaling proteins. Two of the best known of these adapter proteins are insulin receptor substrate-1 (IRS-1) and src-collagen homology proteins (SHC). IRS-1 is especially interesting because it can amplify and diversify effects of IGF-I or insulin by recruitment of a number of intracellular molecules. For example IRS contains about 20 tyrosine phosphorylation sites that when phosphorylated can attract other adapter proteins. Major responses of ligand binding to both IR and IGF-IR include phosphorylation of insulin receptor substrate (IRS) proteins and activation of the mitogen-activated protein kinase(s) (MAPK) as well as activation of phosphatidylinositol-3-kinase (PI3K), which is linked to its cascade of effectors, including serine/tyrosine kinase (Akt), p70s6 protein, and some isoforms of protein kinase C (PKC). PI3K is involved in control of mitogenesis, cellular metabolism, and actin-related cytoskeleton rearrangements (important in cell secretion, mitosis, and ductal morphogenesis). Signaling via the PI3K/Akt pathway is believed critical for cell survival. For example, one of the cellular targets of Akt is a proapoptotic protein (BAD), which promotes cell death when it is bound to the antiapoptotic proteins Bcl_2 or Bcl_{XL}. The continued phosphorylation of BAD by Akt prevents binding and therefore promotes cell survival. This then may explain the long recognized need for addition of insulin (or IGF-I) to maintain mammary tissue in culture. These overlaps in the intracellular signaling cascades explain corresponding similarities in effects of insulin and IGF-I in mammary cells. In contrast, there are at least three other signaling pathways that are preferentially activated by ligand-bound IGF-IR. Examples of IGF-IR and IR signaling and intracellular mediators of ligand binding to these receptors are illustrated in Figure 6.6.

Studies with rodents in which various elements of the IGF-I axis have been deleted confirmed that IGF-I and/or IGF-IR are/is essential for normal mammary development (Kleinberg et al., 2000). Since animals homozygous for absence of the IGF-IR do not survive after parturition, these experiments required the transplantation of the fetal mammary precursor from ~day 18 fetuses into the cleared mammary fat pads of syngeneic hosts. Growth of rudimentary mammary structures from IGF-IR null mice, compared with tissue from wild-type, was minimal. In mice without expression of IGF-I, the development of terminal end buds required replacement with IGF-I. Neither estradiol or GH alone nor the combination had any affect on prepubertal mammary development in these knockout mice. These and related rodent experiments show that in normal peripubertal mammary development, GH acts to bind to GH receptors in the stromal tissue. This is associated with local production of IGF-I, which, in turn, promotes the development of the terminal end buds. In rodents GH was also shown to increase expression of the estrogen receptor in the nuclei of stromal tissue cells, but this effect was not mediated by IGF-I.

The liver is the major source of circulating IGF-I, but a role for locally synthesized IGF-I in the mammary gland is also likely. Assays of samples from rodents, humans, and ruminants indicated that the stromal cells surrounding the epithelium are the source of at least some of the IGF-I detected in the mammary gland. In sheep IGF-I expression is markedly higher in the mammary stroma than in the parenchyma and increases with onset of allometric mammary growth (Hovey et al., 1998). It is difficult to quantify differences in expression of IGF-I between stromal and parenchymal tissue based on assay of whole tissues because the epithelium cannot be completely removed prior to assay (see Fig. 1.1). However, in situ expression data for IGF-I, IGF-II, and IGF-I receptor in the fetal ovine mammary gland strongly indicate that the stromal cells are critical for local IGF-I production. Detection of IGF-I transcripts in enzymatically isolated stromal cells but not in isolated epithelial cells also supports this conclusion. A paracrine model, in which stromal cells adjacent to the epithelium serve as a local source of IGF-I that may then stimulate the growth of the epithelium, is therefore a logical model for IGF-I regulation of mammary development. Certainly, expression of IGF-I mRNA is higher in samples of stromal tissue compared with parenchymal tissue (Fig. 6.7).

In addition to IGF-I there are six IGF-I–binding proteins (IGFBPs) and nine related proteins (IGFBP-rPs) that affect the actions of IGF-I. The IGFBPs are well characterized and bind IGF-I with ~10-fold higher affinity than the IGFBP-rPs. The IGFBPs have several functions including prolonging the half-life of IGF-I, transporting IGF-I from the circulation, and localizing IGF-I to potential target cells. In addition to IGF-I, the mammary gland also synthesizes several IGFBPs (Clemmons, 1998).

In ruminants, the potent effects of insulin, IGF-I, IGF-II, and a natural variant, des (1-3) IGF-I, on mammary cell proliferation has been confirmed by many studies. Reports include experiments with undifferentiated mammary cell organoids

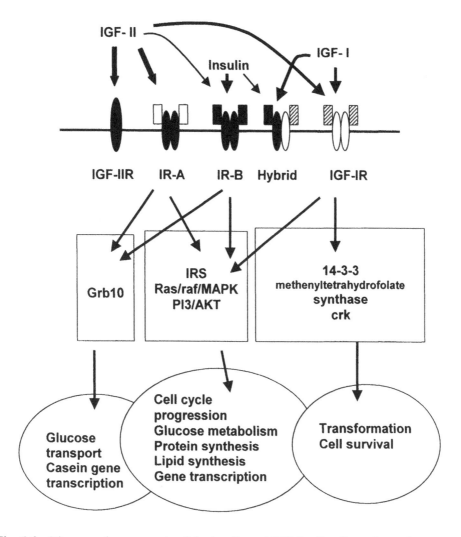

Fig. 6.6 Diagram of components of the insulin and IGF family of proteins and receptors. The upper portion of the figure illustrates binding of IGF-I, IGF-II, and insulin to related receptors. The thickness of the arrows varies with affinity of binding. The center of the figure indicates phosphorylation-signaling cascades stimulated with binding of ligands to different receptors. The bottom of the figure shows cellular events most closely related to a specific receptor-binding event, but the overlap in binding, signaling, and cellular response is also apparent. Adapted from Hadsell and Bonnette, 2000.

from pre- and postpubertal heifers and sheep, pregnant heifers and sheep, and prepartum and lactating cows. In general, des (1-3) IGF-I is the most potent of these agents, followed by IGF-I and IGF-II, but maximal response for each is usually similar. Results demonstrate that the IGFs are very powerful promoters of cell growth with maximal responses often occurring with additions of 20 ng/ml or

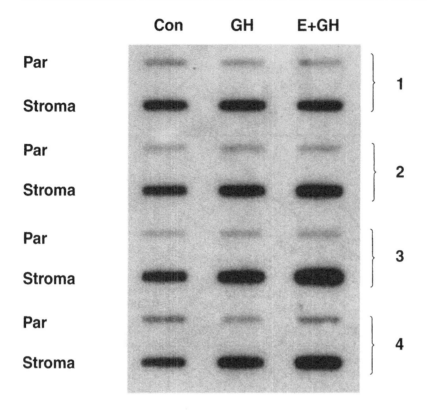

Fig. 6.7 Expression of IGF-I mRNA in parenchymal (Par) and stromal mammary tissue explants from four (1–4) prepubertal heifers. Mammary explants were incubated for 36 hours in the presence of no hormones (Con), GH (1 μg/ml), or estradiol (20 pg/ml) + GH (lane 1, 2, 3). Tissues were removed and processed to isolate total RNA. From each sample 15 μg of RNA was transferred to a membrane and incubated with a radiolabeled cDNA probe specific for IGF-I mRNA. The membrane was washed and exposed to X-ray film. Note the consistent pattern of increased expression of IGF-I mRNA in stromal compared with parenchymal tissue and a tendency for increased expression of IGF-I in stromal tissue incubated with estrogen and GH. Akers, unpublished.

less. Differences in potency of these related molecules reflect differences in affinity for the IGF-I receptor and differences in the capacity of the molecules to associate with IGFBPs. For example, the des- (1-3) IGF-I has limited capacity to bind with IGFBP, but affinity for the IGF-I receptor equals native IGF-I (Fig. 6.8).

Mammary cells derived from prepubertal or nonpregnant, nonlactating heifers, or a line of immortalized bovine mammary epithelial cells (Mac-T cells) grown in collagen gel, show IGF-I inducible secretion of IGFBP-2 and IGFBP-3. Isolated bovine mammary epithelial cells may also secrete smaller amounts of IGFBP-4 and IGFBP-5. But like mammary secretions, IGFBP-2, -3, and -4 are the most prominent. Detection of relatively abundant mRNA transcripts and corresponding proteins for these IGFBPs in heifer mammary tissue suggests the culture system mimics the situation in intact tissue. Much of the mitogenic activity attributed to heifer serum or mammary extracts when applied to cultures of heifer mammary epithelial cell organoids is due to the presence of IGF-I. This is shown by abrogation of the stimulatory effects of serum or mammary extracts by simultaneous addition of antibodies to IGF-I or excess IGFBP-3. Figures 6.8 and 6.9 illustrate the proliferation of bovine mammary cells in response to IGF-I, IGF-II, and des (1-3) IGF-I, as well as the effects of IGFBP-2 and IGFBP-3.

Several lines of evidence support the idea that changes in local tissue secretion of IGF-I and IGFBPs are involved in regulation of mammary growth in the heifer mammary gland. For example ovariectomy of prepubertal heifers at 4 months of age significantly inhibited mammary parenchymal growth and local mammary tissue IGF-I expression at 6 months of age. Corresponding to this, mammary stromal tissue expression of IGF-I mRNA was greater and expression of IGFBP-3 mRNA lower in control compared with ovariectomized heifers. Since increased concentrations of IGFBP-3 are most often associated with blocking the effect of IGF-I, increased IGF-I and lower IGFBP-3 in control compared with ovariectomized heifers correlate with enhanced mammary growth of the intact heifers. Feeding level and GH treatment also affect expression of IGF-I and IGF-binding protein transcripts in the developing heifer mammary gland. Higher concentrations of IGFBP-3 in mammary tissue from highly fed heifers together with reduced mammary tissue sensitivity to IGF-I are possible explanations for the negative effect of high feeding level on mammary growth (Weber et al., 2000b). There are many unknowns, but current evidence suggests that IGF-I and its family of binding proteins are especially important in mammary growth and development. The well-recognized importance of GH in normal mammary development depends on the action of IGF-I.

Although there is little evidence that IGF-I or IGFBPs affect any mammary growth during established lactation, some of these IGFBPs also appear in mammary secretions. It is suggested that IGF-I or IGFBPs in milk or colostrum influence neonatal growth and development. IGFBP-3 is the major binding protein in milk and accounts for approximately 80 percent of the total activity. Interestingly, cows that conceive appear to have higher concentrations in their milk before insemination than those that do not. After pregnancy concentrations of both IGFBP = -2 and -3 fall to very low levels. Concentrations of IGFBP-3 are dramatically higher in mammary secretions near parturition and can equal or exceed concentrations in serum. For example, the concentration in colostrum is ~100-fold higher than in normal milk (Baumrucker and Erondu, 2000).

Epidermal Growth Factor Family

Epidermal growth factor (EGF) was discovered by accident in the 1960s when unexpected biological activity (not attributed to nerve growth factor) occurred following injections of extracts of murine salivary glands into test animals. These effects were associated with precocious eyelid opening and tooth eruption. Subsequent studies showed that preparations of EGF stimulated the proliferation of isolated epidermal cells and tissue explants, hence the name epidermal growth

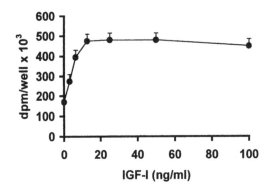

A

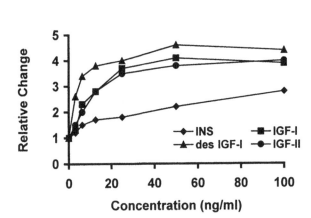

B

Fig. 6.8 Proliferation of bovine mammary epithelial cell organoids from prepubertal heifers in response to the IGF-I (**A**) and relative response to IGF-I, IGF-II, insulin (INS), or des IGF-I (**B**). Organoids were cultured inside collagen gels, and growth measured by incorporation of ^{3}H-thymidine. Data adapted from Purup et al., 2000a, or Weber et al., 1999.

factor. Study of a human epidermal cancer cell line (A-431) was fundamental because these cells greatly overexpressed receptors for EGF (~3 × 10^6 receptors per cell). This property allowed for the isolation and eventual structural characterization of the receptor. The availability of relatively large quantities of the receptor also led to the discovery that the addition of EGF to isolated membranes stimulated the phosphorylation of both endogenous membrane proteins as well as many exogenously added proteins. It was subsequently shown that the receptor protein was a 170 kDa transmembrane glycoprotein whose external domain formed the binding site for EGF and whose cytoplasmic domain possessed tyrosine kinase activity. EGF bound to the external portion of the receptor activates

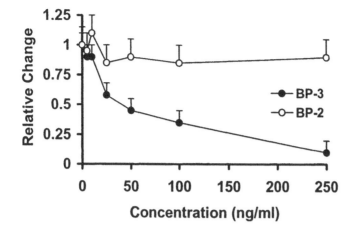

A

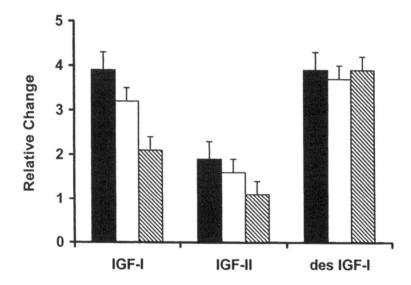

B   IGF-I   IGF-II   des IGF-I

Fig. 6.9   Effect of IGFBP-2 or IGFBP-3 on DNA synthesis in bovine mammary organoids cultured in collagen gels is shown. In **A** increasing amounts of IGFBP-2 or IGFBP-3 were added to cultures containing 6.25 ng/ml of IGF-1. Concentrations of IGFBP-3 > 10 ng/ml reduced the effect of IGF-I, but addition of IGFBP-2 in concentrations up to 250 ng/ml had no effect on IGF-I–stimulated cell growth. **B** illustrates the effect of IGFBP-3 (0, 25, 50 ng/ml, *solid, open, and patterned bars, respectively*) added to organoids cultured inside collagen gels in the presence of 6.25 ng/ml IGF-I, IGF-II, or des IGF-I. Growth was measured by incorporation of ^{3}H-thymidine. Note the lack of effect of IGFBP-3 on des IGF-I stimulation of DNA synthesis. Data adapted from Purup et al., 2000a.

the cytoplasmic or catalytic domain of the receptor to autophosphorylate and phosphorylate cytoplasmic substrates essential for EGF action. These observations were important because they were among the first to directly link ligand binding, receptor activation, and phosphorylation with general mechanisms of action for many growth factors.

This family of proteins contains at least 10 members, including EGF, heparin-binding EGF, TGF-α, amphiregulin, and several heregulins. Each of these structurally similar proteins acts by binding to one of several related membrane receptors called the type I receptor tyrosine kinases (RTKs) or the ErbB family of receptors. The usual EGF receptor is called ErbB-1, but at least four variants are known. Certain ligands and receptors of the family, ErbB-2 for example, contribute to the aggressive phenotype of some human breast carcinomas and related poor prognosis. A homologue of EGF, transforming growth factor α (TGF-α) was first isolated from the medium of oncogene-transformed cells, and a transforming avian retrovirus was subsequently shown to code for an abbreviated form of the EGF receptor. EGF and TGF-α are highly expressed in early embryonic development. Amphiregulin and HB-EGF are secreted heparin-binding growth factors.

There is very little known about a possible mitogenic role for these proteins in ruminant mammogenesis (Plaut, 1993). Specific EGF receptors are found in mammary tissue from sheep and cows, and expression of EGF mRNA has been shown for bovine mammary tissue (Koff and Plaut, 1995). Ligand-binding assays using either radiolabeled EGF or TGF-α show a single class of high-affinity binding sites in mammary tissue of sheep and cows during gestation and into lactation. However, the number of receptors is greater during mid- than during late pregnancy or lactation. Expression of TGF-α occurs in bovine, and amphiregulin occurs in ovine mammary tissue. Addition of either TGF-α or EGF stimulates DNA synthesis in explants from heifers near the middle of gestation and in epithelial cells from heifers or pregnant cows and sheep. Figure 6.10 shows the effect of EGF on DNA synthesis in mammary explants prepared from tissue taken from heifers near the middle of gestation, either before or after xenotransplantation into nude mice. For freshly prepared explants, concentrations of EGF of less than 10 nM only stimulated DNA synthesis after 2 days in culture. For xenotransplanted tissues, priming of mice with estrogen and progesterone for 10 days increased the sensitivity of the tissue to EGF when explants were removed and tested in culture. Interestingly, treatment of the mice with hydrocortisone or Prl for the final 2 days of transplantation slightly reduced the subsequent response to EGF, but treatment with hydrocortisone + Prl blocked any effect of EGF on DNA synthesis. This response suggests that mammogenic effects of EGF may be more important for mammary ductular development than for lobulo-alveolar formation in cattle. Failure of EGF to stimulate DNA synthesis in mice given hydrocortisone and Prl suggests that differentiation of the alveolar cells is incompatible with EGF-induced growth.

Intramammary infusion of EGF stimulated DNA synthesis in udders of pregnant heifers, but it is not clear if this represents augmentation of a normal response to naturally occurring EGF in the bovine mammary gland or a pharmacological

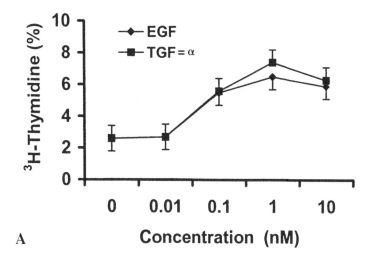

A

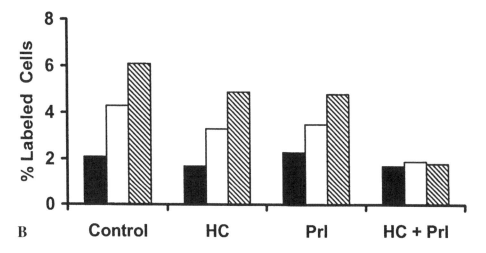

B

Fig. 6.10 **A** shows autoradiographic analysis of EGF- or TGF-α–stimulated DNA synthesis in bovine mammary tissue explants from Holstein heifers near the middle of gestation. Tissues were incubated for 2 days in the presence of the growth factors. DNA synthesis was measured during the last 6 hours of culture by measurement of tritiated thymidine incorporation. **B** shows the effect of hydrocortisone (HC) and Prl on EGF-induced DNA synthesis in bovine mammary alveolar epithelial cells for tissue explants maintained in athymic nude mice. Mice were primed for 10 days with estrogen and progesterone, then treated for 2 days with placebo (control), HC, Prl, or both. Xenografted tissues were removed and cultured for 2 days with no EGF (*solid bars*), 0.1 nM EGF (*open bars*), or 1 nM EGF (*patterned bars*). Note the lack of effect of EGF for tissues when mice were treated with HC + Prl. Data adapted from Sheffield, 1998.

effect. Neither EGF nor TGF-α stimulated DNA synthesis in the bovine Mac-T mammary epithelial cell line. It is possible that this is because the cells were derived from a lactating cow or that the cells were grown in monolayer cultures on tissue culture plastic for these studies. Positive effects on explants and for primary cell organoids may reflect an improved capacity of the cells to respond if cultured with extracellular matrix proteins (Woodward et al., 2000). For example, incubating the murine mammary cells on laminin, collagen, or fibronectin impacts response of the cells to IGF-I and EGF. The presence of mRNA for EGF, TGF-α, and amphiregulin suggests that these growth factors may be important in mammogenesis in pubertal heifers and during pregnancy. Unlike EGF and TGF-α, amphiregulin binds to heparin so that its action is blocked by heparin because it is sequestered from interacting with the EGF receptor. Studies with ovine mammary cells show that DNA synthesis is inhibited by heparin, but the inhibition can be overcome with addition of TGF-α. The effect, however, varies with stage of gestation. Unlike EGF or TGF-α, human amphiregulin has two nuclear targeting sequences and can be detected by immunocytochemical techniques in the nuclei of test cells. Site-directed mutagenesis of this region of the molecule also abolishes the mitogenic activity of the molecule. This suggests that nuclear localization may be important in the growth-stimulating effect of amphiregulin, but this further indicates that there are differences in mechanisms of action compared with EGF or TGF-α. Arguing against the idea that the EGF family is essential for mammary development, transgenic mice that lack expression of TGF-α, EGF, or both can still lactate normally. In these cases, it could be that other EGF-like molecules can substitute for the missing proteins, but this seems unlikely.

For tissues that synthesize EGF or its related ligands, these growth factors are made as transmembrane glycoproteins. The EGF-like sequences are external to the plasma membrane. Requirements for secretion are not completely understood but involve the action of a protease that cleaves the membrane-bound protein for release of the EGF (or relative) into the interstitial fluid. Induced overexpression of TGF-α under control of mammary promoters in rodents leads to the appearance of mammary tumors. This fits with measurements showing that some spontaneous human mammary tumors also express large amounts of TGF-α or EGF receptor. It is also important to remember that responses of cells or tissues in culture may not necessarily reflect the response of mammary tissue in the complex environment of the intact animal. Presence of EGF receptors and ligands and demonstrated response of ruminant mammary cells to EGF or its relatives support a role for these growth factors in mammogenesis, but much more information is needed to determine exact roles. In short it is not known if the EGF family of proteins and receptors is largely permissive or if it has an essential direct impact in ruminant mammary development.

Fibroblast Growth Factors

Fibroblast growth factors (FGFs) constitute a family of 20 small peptide growth factors that share a highly homologous core of 140 amino acids and strong affinity for heparin and heparin-like glycosaminoglycans (HLGAGs) of the extracel-lular matrix (ECM). FGF signaling and interactions with the ECM have attracted the attention of numerous cancer researchers because of the potential role of the FGFs in promoting the progression of some tumors from a hormone-dependent to a hormone-independent pattern of growth. Others are attracted to FGF studies from observations that development of many cancers positively correlate with local tissue secretion of proteases and changes in pH that might act to change con-centrations of biologically available FGFs in the local tissue environment. Both FGF-1 (also known as acidic FGF or aFGF) and FGF-2 (also know as basic FGF or bFGF) were first identified from extracts of bovine pituitary glands based on the capacity of the proteins to stimulate DNA synthesis in cultured fibroblasts. Members of this family of growth factors are linked by structural similarities and capacity to bind heparin or HLGAGs, not by the specificity of their growth-stim-ulating activity. By convention they continued to be designated as FGFs despite the fact that not all of the proteins stimulate proliferation of fibroblasts. Consequently, several members of the FGF family have emerged as stroma-derived mitogens, which may act in a paracrine manner to locally influence mam-mary epithelial cell proliferation as well as mammary gland morphogenesis (Hovey et al., 1999; Powers et al., 2000).

Although the FGFs function after they appear in the extracellular environment via binding to high-affinity cell surface receptors, neither FGF-1 nor FGF-2 are syn-thesized with a leader peptide sequence. As described in relation to secretion of milk proteins by the mammary alveolar cells (Fig. 3.8), the leader sequence is a strand of hydrophobic amino acids at the amino terminal of the newly synthesized peptide that serves to control the secretion destination of the protein. The leader peptide is recognized by a signal recognition particle (SRP), which temporarily halts transla-tion and serves to transport the translation complex to the endoplasmic reticulum (ER). At this point protein synthesis resumes, and the nascent peptide chain is vec-tored into the cisternal space of the ER and subsequently passed to the Golgi for packaging into secretory vesicles for secretion from the cell. This feature has attracted cell biologists to the study of these FGF variants to decipher this secretion mechanism. Because FGFs are involved in wound healing, it has been suggested that mechanical damage provides a mechanism for release of FGF from endothelial cells, but such a mechanism would seemingly lack the regulation necessary for secretion of FGFs in normal mammary tissue development.

FGF-1 likely functions as a paracrine mitogen for mammary epithelial cells. For example, transgenic mouse experiments that targeted overexpression of a defective FGF receptor to the mammary gland showed that lobulo-alveolar development was markedly impaired. Of the various rat mammary cell types, fibroblasts express the greatest level of FGF-1 mRNA in vitro. Appearance of mRNA and protein within both the intact mouse mammary gland and epithelial-cleared mammary fat pad strongly supports the idea that it is stromal in origin. At least three of the known FGF variants (FGF-1, FGF-2, and FGF-7, also known as keratinocyte growth fac-tor [KGF]) are proposed to be involved in ruminant mammary development. These

FGFs and their receptors are expressed during mammogenesis, lactation, and mammary involution. The highest levels of expression are in glands of virgin heifers and in primiparous heifers during involution.

Although FGF-2 mRNA expression is greatest in the stromal tissue, immunocytochemical studies show that the FGF-2 protein associates with myoepithelial cells. This distribution may reflect the high affinity of FGF-2 for specific components of the ECM, further supporting its proposed role as a paracrine/autocrine mitogen for myoepithelial cells. Synthesis of FGF-2 in the mouse mammary fat pad is hormonally regulated based on observations that expression is greatest during late pregnancy in correspondence with the appearance of higher tissue concentrations of the protein at this time. Expression is also increased in the bovine mammary gland in late gestation. Interaction between epithelium and the surrounding stroma likely influences paracrine FGF-2 expression. In rodents and ruminants expression is greater in the stroma adjacent to the developing parenchymal tissue. The capacity of either FGF-1 or FGF-2 to stimulate DNA synthesis in bovine mammary tissue organoids in vitro is show in Figure 6.11. Interestingly, the effect of FGF-2 is relatively small and biphasic in comparison with FGF-1.

FGF-7 is also likely a paracrine mediator of hormone action for mammary tissue. In the fetal mammary gland, it is localized to the mesenchymal cells surrounding the developing mammary bud. In the postnatal period ovine mammary stromal cells continue to express FGF-7 mRNA in vitro and in vivo. Two transcripts of 2.4 and 1.5 kb found in the mammary fat pad are associated with expression in fractions of fibroblasts versus adipocytes, respectively. In vivo expression of FGF-7 in ovine stromal tissue is suppressed by exogenous estradiol or by addition of glucocorticoids in vitro. For the peripubertal ewe, it is proposed that local stromal synthesis of FGF-7 is controlled by positive feedback from the adjacent epithelial cells. To illustrate, FGF-7 expression was higher in mammary stromal tissue adjacent to the developing parenchymal tissue than in mammary stromal tissue taken from the contralateral mammary gland in which the epithelium had been surgically removed prior to the onset of allometric mammary growth (Hovey et al., 2001). It is possible that FGF-7 accounts for some of the species differences in parenchymal morphology. In humans and ruminants, peripubertal mammary parenchymal development follows a lobulated pattern, in the presence of a dense, abundant fibroblastic stroma, distinct from that of rodents. In the virgin rodent mammary gland, these dense stromal elements are greatly reduced, but adipocytes are abundant. Correspondingly, expression of FGF-7 is reduced compared with ruminant mammary tissue. Finally, treatment of mice with exogenous FGF-7 can induce a more ruminant-like pattern of growth in association with changes in the stromal tissue matrix. Figure 6.12 shows the pattern of expression for FGF-7 in ovine mammary tissue during the onset of allometric mammary development. As is the case with the EGF-related peptides and receptors, it is not possible at this time to develop an encompassing global view of how the activities of the FGFs and related receptors might interact with the classic mammogenic hormones to control normal mammary development.

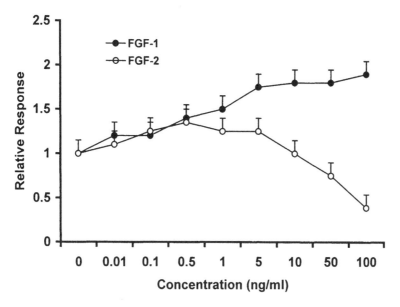

Fig. 6.11 Effect of bovine FGF-1 and FGF-2 on DNA synthesis in prepubertal bovine mammary organoids cultured inside collagen gels is shown. The value 1 represents the growth response for cells cultured in basal medium (control). Data adapted from Purup et al., 2000b.

Transforming Growth Factor -β

The TGF-β group of proteins is made up of at least five multifunctional proteins that have functions ranging from modification of the ECM, to induction of differentiation of target cells, to stimulation of proliferation in multiple cells types and tissues. Three of the variants, TGF-β1, TGF-β2, and TGF-β3, stimulate connective tissue formation and are chemotactic for fibroblasts. They can indirectly promote proliferation of mesenchymal cells but can inhibit growth of epithelial cells in vivo and in vitro. These varied effects of TGF-βs suggest that they are likely important in mammary development and function. TGF-β1 is the best described of these proteins related to mammary function. Blood platelets provide the most concentrated source of TGF-β1, but it is believed to be produced by nearly every cell type in the body. Biologically active TGF-β1 is a 25 kDa disulphide-linked homodimer. When secreted, it is bound to a large 75-kDa glycoprotein called the latency-associated peptide (LAP). Activation of the latent form by proteases, alkalinization, or chaotropic agents is necessary for TGF-β1 to bind to its receptor, so control of this reaction is an important regulator of TGF-β1 action. Unregulated mammary epithelial cell proliferation is obviously an undesirable trait of mammary tumor formation, so it should not be surprising that actively growing cells must be controlled to prevent hyperplasia. Most of the mammary-associated effects of TGF-βs are inhibitory.

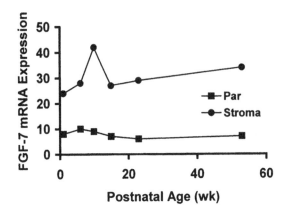

Fig. 6.12 Expression of FGF-7 mRNA in ovine mammary tissue from virgin ewes is shown. Average relative expression is approximately four-fold greater in mammary stromal tissue. Data are given as arbitrary density units corrected for RNA loading. Data adapted from Hovey et al., 2001.

Effects of TGF-β are mediated by binding to specific cell surface receptors (designated type I, II, or III receptors) present on most cell types. The type I and type II receptors are directly involved in signal transduction, while the type III receptor is thought to enhance binding of TGF-β to one of the other receptor subtypes. The ductal epithelium of heifer mammary cells shows extensive presence of type I and type II receptors by immunocytochemical localization of antibody to the receptors

Some of the best evidence of the importance of TGF-β1 in normal mammary ductular development has come from mouse studies in which small plastic implants containing small amounts of TGF-β1 were shown to reversibly inhibit the growth of end buds in situ (Daniel et al., 1996). Application of exogenous TGF-β inhibits duct cell proliferation within hours and over several days, inducing localized ECM synthesis. In the mouse, overexpression of TGF-β transgenes at the time of puberty markedly reduces the rate of duct development as well as the degree of ductular tree expansion. Expression during pregnancy impairs lobulo-alveolar development and therefore lactation. A working model from rodent studies is that TGF-β1 acts to stabilize ductular structures in a concentration-dependent manner to prevent duct entanglement by altering ECM formation during the period of rapid duct elongation. Such a model then would depend on the relaxation of these restraints for formation of side branches and alveolar budding (Smith, 1996).

The role of TGF-β1 in bovine mammary development is unknown, but concentrations of TGF-β1 in serum ranged from 7 to 30 ng/ml, and receptors for TGF-β1 in bovine mammary membranes are increased during the peripubertal period, corresponding with rapid mammary development. TGF-β1 is also expressed in the mammary tissue of virgin heifers. In vitro TGF-β1 produces a biphasic effect on proliferation of bovine mammary epithelial cell organoids from prepubertal heifers.

Concentrations up to 500 pg/ml stimulate proliferation, but concentrations greater than 1 ng/ml were inhibitory. Maximal inhibition occurred at 5 ng/ml, and this was associated with ~85 percent inhibition of the stimulatory effect in the presence of 5 percent fetal calf serum. Addition of an antibody to TGF-β blocked inhibition of cell growth at the higher concentrations, indicating the specific nature of the inhibition. Related studies show that TGF-β1 inhibits the proliferation due to addition of IGF-I, IGF-II, des (1-3) IGF-I, EGF, or amphiregulin. TGF-β1 also affects the morphology of bovine mammary organoids. Some of these responses are shown in Figure 6.13. Certainly, the possibility that IGFBP-3 (or fragments) might have IGF-I receptor–independent actions in mammary cells, via binding to the type V TGF-β receptor, coupled with TGF-β induction of IGFBP-3, makes for an intriguing overlap between the growth-stimulating actions of the IGF-I axis and the inhibitory effects of the TGF-β family of molecules.

Hepatocyte Growth Factor

The first evidence for hepatocyte growth factor (HGF) involvement in mammary development was derived from observations on the effect of conditioned media from a line of human fibroblasts (MRC 5 cells) on growth and morphogenesis of human mammary cells in culture. Specifically, cell colonies changed from contiguous sheets of cells to loose clusters of cells without the usual junctional complexes. Because of this effect, the substance in the conditioned medial was called *scatter factor*. Isolation and structural characterization of the protein showed that it was identical to HGF. Thus, HGF stimulates not only hepatocytes but also many other epithelial cell types. When added to mammary cells cultured inside collagen gels, HGF induces growth and the appearance of branching tubules reminiscent of mammary ducts.

HGF effects depend on binding to a surface receptor tyrosine kinase encoded by the c-met protooncogene (cMet). The receptor is a 190-kDa heterodimeric protein with a 50-kDa α- and a 145-kDa β-subunit. The α-subunit and the NH_2 terminal fragment of the β-subunit are exposed on the cell surface and are involved in binding with HGF. The COOH terminal fragment of the β-subunit passes into the cytoplasm and contains the tyrosine kinase domains and the associated docking sites for intracellular regulators of the kinase. The Ras-MAP kinase-signaling pathway mediates the mitogenic effects of HGF. The morphological effects to form tubular structures involve activation of Gab 1, Stat-3, and SMAD 1 as intracellular mediators (Kamalati et al., 1999).

HGF mRNA localizes to stromal cells within the normal mouse mammary gland, but there are also indications for multicellular upregulated expression of HGF in human breast cancer. During postnatal development of the rat and ovine mammary gland, expression of HGF is greatest in mature virgin and involuting states, supporting a role for HGF in promotion of ductal elongation and branching as well as tissue remodeling between lactations. HGF actions are hormone regulated, since expression of the HGF receptor is increased by prolactin and glucocorticoids. These observations support a role for HGF derived from constituents

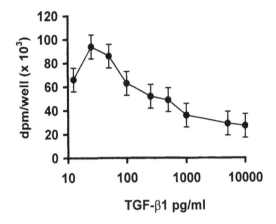

A

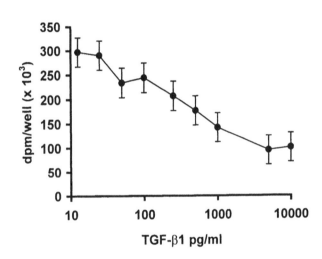

B

Fig. 6.13 Effect of TGF-β1 on DNA synthesis in prepubertal bovine mammary organoids is shown. **A** shows the effect of TGF-β1 added to basal (control) medium, and **B** the effect of TGF-β1 added to IGF-I (50 ng/ml). TGF-β1 is a potent inhibitor of DNA synthesis. Data adapted from Purup et al., 2000a, and Ellis et al., 2000.

of the stroma in parenchymal morphogenesis, but data specific to ruminants are limited (Hovey et al., 1999).

Parathyroid Hormone–Related Peptide

Parathyroid hormone-related protein (PTHrP) was initially described as a tumor product involved in a clinical syndrome called *humoral hypercalcemia of malignancy*. Patients with the syndrome exhibit symptoms of hypercalcemia and hypophosphatemia that are similar to effects seen in primary hyperparathyroidism. For this reason it was often assumed that effects were caused by tumors producing parathyroid hormone. Isolation of proteins from the tumors of these patients showed that this was not the case. Cloning of the PTHrP gene led to a variety of comparative studies and subsequently to data showing that PTHrP and its receptor are expressed in many different tissues and that they make up one of the earliest hormone-receptor pairs appearing during development (i.e., it is expressed at the late morula stage and thereafter). These observations suggested that PTHrP is important in organogenesis and development generally. Initial attempts to link gross expression of PTHrP with mammary development failed because PTHrP gene knockout mice die at birth because of skeletal abnormalities. Subsequent experiments allowed the development of mice devoid of PTHrP in all tissues except the skeletal system. These mice had multiple postnatal developmental problems in ectodermally derived structures (e.g., the integumentary system, teeth, sebaceous gland, etc.), but effects were especially dramatic in the mammary gland. The animals had normal mammary fat pads but no epithelial components. These data indicate a dramatic involvement of PTHrP in regulation of the epithelial-mesenchymal interactions necessary for fetal mammary development. PTHrP and its receptor may also be involved in postnatal mammary development since expression is most intense in the regions surrounding the terminal end buds, but there was little or no expression in mature mammary ducts. A current hypothesis is that PTHrP interacts with the PTH/PTHrP receptor on mammary stromal cells and that this interaction is critical to support normal mammary ductular morphogenesis (Dunbar et al., 1998). The observation that mammary stromal cells express the receptor but not PTHrP supports this idea. In parallel, isolated mammary epithelial cells express the PTHrP mRNA but not mRNA for the receptor. Last, mammary epithelium from PTHrP receptor knockout embryos can survive and begin ductal morphogenesis if combined with wild-type mesenchymal tissue and transplanted beneath the kidney capsule of athymic nude mice (Dunbar and Wysolmerski, 1999). The role of PTHrP or its receptor in ruminant mammary development is unknown.

Other Stroma-Derived Growth Regulators

In addition to factors described above, the list of stroma-derived growth regulatory molecules continues to expand although the precise physiological function(s) for many of these is (are) unknown. Among other such molecules, Wnt gene family members are expressed by constituents of the mouse mammary fat pad in vivo (Wnt 2, 5a, and 6) and depend on ovarian hormone regulation. Epimorphin, a stroma-derived polypeptide with morphogenic effects on epithelium in vitro, has been described. Similarly, neu-differentiation factor is expressed by the perialveolar stroma during pregnancy and induces lobulo-alveolar growth and milk protein synthesis in vitro. Connective tissue growth factor is among other factors that are expressed by the mammary stroma in disease states. Finally, stroma-derived proteases are also certainly important for local effects on growth and morphogenesis.

For example, mammary stroma-derived cathepsin D cleaves prolactin, while stromelysin-1 and –3, and gelatinase, are expressed by mammary stromal cells and influence parenchymal morphogenesis and tissue remodeling. As the number of paracrine-acting factors continues to unfold, emphasis needs to be placed on their precise function during the course of mammary epithelial growth and neoplasia, with emphasis especially on their roles in mammary morphogenesis in economically important dairy species.

In summary, as Tucker (2000) admonished in a recent review, "A plethora of growth factors and binding proteins have been implicated in mammogenesis. But, why are there so many growth factors? Surely they all don't serve the same function?" Certainly, it seems irrational from an evolutionary viewpoint that the cells and tissue would invest the metabolic resources to synthesize and secrete a myriad of agents for the sake of producing so many redundant controls of mammary development and function.

Chapter 7
Endocrine, Growth Factor, and Neural Regulation of Mammary Function

Just as the endocrine system is critical for mammary growth and development, it also stimulates the final phases of lobulo-alveolar differentiation and coordinates onset of milk secretion at the time of parturition. Once lactation is established, interactions between the endocrine and nervous system provide the signals for continued milk synthesis, secretion, and milk removal.

Endocrine Regulation of Lactogenesis

Now confirmed in numerous species, classic mammary explant culture studies demonstrated that the major positive regulators of differentiation of the secretory cells are glucocorticoids and prolactin. Mammary tissue explants from pregnant or steroid hormone–primed donors exhibit both biochemical and structural differentiation when incubated in culture medium containing a combination of insulin, glucocorticoids, and Prl. While it is widely believed that insulin is necessary for mammary tissue maintenance in culture, species variation in insulin sensitivity of the mammary gland in intact animals casts some doubt on this belief. Recent data also support the idea that insulin-mediated effects on mammary cells in culture may actually represent effects more appropriately ascribed to IGF-I. This is because mammary epithelial cells have specific IGF-I receptors, and insulin (especially at higher concentrations typical of culture experiments) is likely to bind to the IGF-I receptor. In general terms glucocorticoids are most closely associated with development of rough endoplasmic reticulum and Prl with maturation of the Golgi apparatus and appearance of secretory vesicles. Prl added to mammary explant cultures incubated with insulin and glucocorticoids dramatically increases the de novo synthesis and secretion of α-lactalbumin and caseins. Figure 7.1 illustrates the effects of glucocorticoids and Prl on the capacity of mammary explants from nonlactating, pregnant cows to secrete α-lactalbumin. Since production of α-lactalbumin is closely tied with normal milk secretion and it is a specific milk protein, changes in its production provide a good reference for differentiation of the alveolar secretory cells. These data show that Prl is

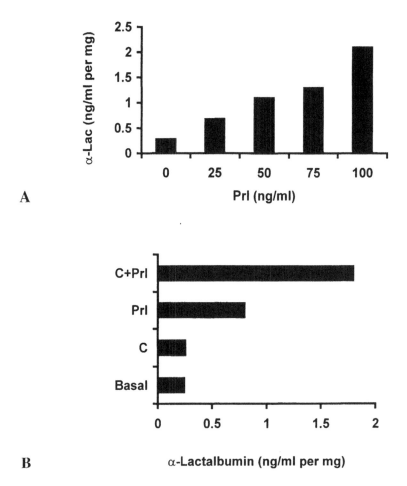

Fig. 7.1 Effect of lactogenic hormones on α-lactalbumin (α-Lac) secretion by bovine mammary explants taken from multiparous, nonlactating cows is illustrated. **A** demonstrates a concentration-dependent effect of Prl on α-lactalbumin secretion. Explants were incubated for 48 hours in basal medium also supplemented with insulin (5 μg/ml), triiodothyronine (0.65 ng/ml), and cortisol (500 ng/ml). **B** shows the additive effect of cortisol (C) on Prl stimulation of α-lactalbumin secretion. Explants were also incubated with triiodothyronine and insulin as in **A**. Data adapted from Goodman et al., 1983.

a major stimulator of α-lactalbumin secretion but that the capacity of the tissue for α-lactalbumin secretion is improved with the addition of hydrocortisone and that simultaneous addition of estradiol and triiodothyronine further improves the response. In keeping with the in vivo effects of progesterone to inhibit premature onset of lactogenesis, addition of exogenous progesterone reduces Prl-induced α-lactalbumin secretion (Fig. 7.2).

Application of molecular techniques to mammary gland biology has served to solidify the idea that Prl and glucocorticoids are primary stimulators of mammary cell differentiation. For example, both Prl and glucocorticoid response elements

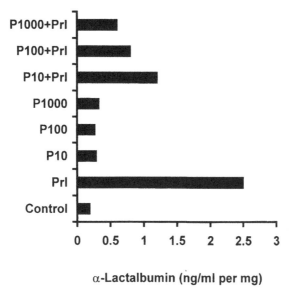

A

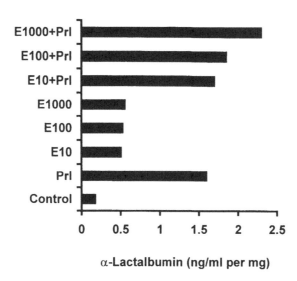

B

Fig. 7.2   Effect of lactogenic hormones on α-lactalbumin (α-Lac) secretion by
bovine mammary explants taken from multiparous, nonlactating cows is illustrated.
A demonstrates a concentration-dependent, inhibitory effect of progesterone (P) at 10,
100, or 1000 ng/ml on Prl (100 ng/ml) -induced secretion of α-lactalbumin. **B** illustrates the small stimulatory effect of estradiol (E) at 10, 100, or 1000 ng/ml alone as
well as the effect of E to enhance the response to Prl (100 ng/ml). Explants were incubated for 48 hours in basal medium also supplemented with insulin (5 μg/ml), triiodothyronine (0.65 ng/ml), and cortisol (500 ng/ml). Data adapted from Goodman et
al., 1983.

are found within the promoter regions of the genes for several mammary-specific milk proteins. Similarly, induction of both mRNA and specific milk proteins in response to addition of Prl or glucocorticoids in isolated mammary epithelial cells confirms the importance of these hormones in lactogenesis and provides details for mechanisms of action of these hormones in control of milk protein gene expression (Rosen et al., 1999).

Periparturient Endocrine Profiles

Just as measurements of circulating concentrations of various hormones aided elucidation of hormones correlated with the changes in mammogenesis, measurement of changes in serum hormones around the time of parturition helped confirm the importance of changing secretion of Prl, glucocorticoids, and progesterone in control of lactogenesis.

On the other hand, despite the continuous presence of both Prl and glucocorticoids in the circulation during gestation, stage II of lactogenesis is held in check until the period just prior to parturition. Essentially, the combined effects of positive stimulators (Prl, glucocorticoids, growth hormone, and estradiol) and the reduced negative influence of progesterone interact to determine the timing of the onset of copious milk secretion. Many studies have reported changes in blood concentrations of these hormones in correspondence with parturition. In dairy ruminants, for example, there are consistently large increases in concentrations of serum prolactin in the period around parturition and acute secretion of glucocorticoids in closer association with the actual birth process. Concentrations of estradiol progressively increase during late gestation to a maximum within a few days of parturition. In contrast, in association with luteolysis of the corpus luteum, concentrations of progesterone abruptly decline 3 to 4 days prior to parturition. Changes in other hormones (prostaglandin $F_{2\alpha}$ also increases at parturition) as well as changes in hormone and growth factor concentrations in mammary secretions may also serve to regulate stage II lactogenesis. There are also marked changes in concentrations of IGF-binding proteins during the periparturient period. However, it remains to be seen if IGF-I or IGF-II is acutely involved in lactogenesis. It is also possible that such changes correspond with a decrease in mammary cell proliferation with the onset of lactation, given that IGF-I is a potent stimulator of mammary cell proliferation. In rodents at least, there is also evidence that EGF may act to modify the ability of Prl to stimulate synthesis and secretion of milk proteins. Neither should it be forgotten that many of the growth factors and other biologically active substances in milk, especially during the early postpartum period, might have evolved for a role in gastrointestinal tract physiology of the neonate independent of possible roles in maternal mammary physiology. Changes in serum concentration of several relevant hormones around the time of parturition are shown in Figure 7.3.

Although progesterone can inhibit lactogenesis, simple removal of progesterone does not necessarily induce onset of lactation. In many species, lactogene-

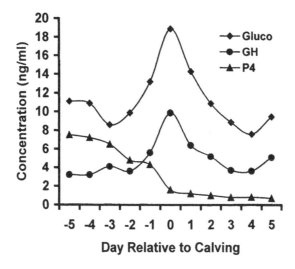

A

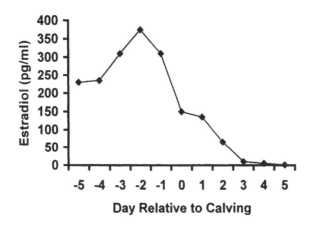

B

Fig. 7.3   Changes in serum concentrations of mammogenic and lactogenic hormones at the time of parturition in cattle are shown. Concentrations of GH and glucocorticoids increase near calving but decline to basal concentrations shortly thereafter. Progesterone (P4) declines quickly over several days just before calving (**A**). Estradiol concentrations also rapidly decrease from very high concentrations reached in the week before parturition (**B**). Data adapted from Akers et al., 1981a, and Tucker, 1994.

sis is well under way before the decline in progesterone near parturition. Furthermore, prepartum milking causes premature differentiation of the mammary epithelium and subsequent evolution of mammary secretions of essentially normal composition often several weeks before parturition. Thus, differentiation of the alveolar epithelium and onset of milk synthesis and secretion is possible despite presence of high progesterone concentrations in the blood (Akers et al., 1977; Akers and Heald, 1978).

Table 7.1 Changes in lactose synthesis, α-lactalbumin, RNA, and DNA in mammary gland of cows before and after parturition and with treatment with CB154 to suppress Prl secretion

	Treatment			
	Prepartum	*Postpartum*	*CB154*	*CB154 + Prl*
Lactose synthesis (µg/h/100 mg)	39 ± 63	552 ± 70	327 ± 63	628 ± 81
α-Lactalbumin (µg/mg/cytosol protein)	1.7 ± .04	5.4 ± 0.4	2.8 ± 0.4	6.8 ± 0.5
RNA (g)	23.6 ± 2.1	87.2 ± 16.8	56 ± 7.9	91.5 ± 12.8
DNA (g)	27.9 ± 2.9	46 ± 3.8	40.1 ± 3.8	42.2 ± 3.8

Source. Adapted from Akers et al., 1981b.
Note. Control animals were killed 10 days before or after parturition. CB154 was administered for 12 days prior to expected parturition through parturition. Animals given Prl were infused continuously for 6 days immediately before parturition.

Essential Roles of Prl, Glucocorticoids, and Progesterone

Some of the best evidence for the importance of increased periparturient secretion of Prl in stage II lactogenesis has come from experiments in which the administration of a dopamine agonist has been used to inhibit Prl secretion and correspondingly impair lactation. In ruminants where postpartum milking continues, administration of the dopamine agonist α-bromoergocryptine (CB154) reduced basal Prl concentration about 80 percent and prevented the usual periparturient rise as well as the milking-induced Prl rise during the first week postpartum. Differences in Prl secretion in control and CB154-treated cows are shown in Figure 7.4. Milk production was reduced 45 percent during the first 10 days postpartum. Lost milk production was associated with reduced synthesis of α-lactalbumin, lactose, and fatty acids, as well as impaired structural differentiation of the mammary secretory cells. Selected effects are summarized in Table 7.1. Cows treated with exogenous Prl in addition to the agonist (to replace the periparturient surge in prolactin) showed no loss of milk production or effects on milk component biosynthesis or alveolar cell differentiation. An effect of Prl suppression and replacement during the periparturient period on milk production in multiparous cows is shown in Figure 7.5. Clearly, Prl is important in mammary cell differentiation and lactogenesis (Akers et al., 1981a, b).

That the periparturient period is critical can also be gleaned from experiments in which intramammary infusion of the microtubule inhibitor colchicine during the week prior to parturition also markedly inhibited subsequent structural and biochemical differentiation of the secretory epithelium. This suggests that in vivo mammary cell differentiation normally occurs in a relatively short period and that disruption of the process during this time may have long-term consequences on milk production.

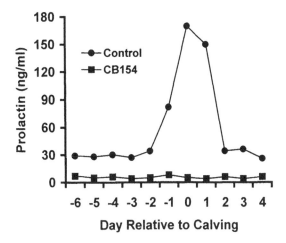

Fig. 7.4 Changes in serum Prl in cows given ergocryptine (CB154) at the time of calving are shown. Note that the usual surge in Prl secretion at calving is blocked and that average Prl concentration before and after calving is reduced, but secretion of Prl is not completely suppressed. Adapted from Akers et al., 1981a.

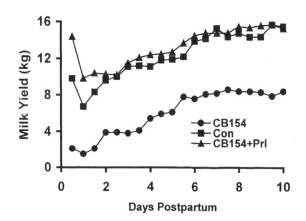

Fig. 7.5 Milk production per milking in control, CB154-treated, and CB154 + Prl–treated cows is shown. Note the marked decrease in milk yield of cows given CB154 but that milk yield is restored to normal for cows also given Prl. Data adapted from Akers et al., 1981a.

Aside from circumstantial evidence related to changes in circulating hormone concentrations, there is a marked increase in numbers of mammary cell receptors for Prl, IGF-I, and cortisol during late gestation. It is also relevant that progesterone receptor concentration is correspondingly reduced with the onset of lactation. Thus, simultaneous, coordinated changes in circulating hormones and receptors appear to regulate the timing of lactogenesis.

Data from a variety of tissue culture experiments also strongly support these hormones playing a role in lactogenesis. For example, additions of estradiol or cortisol markedly enhance Prl-induced secretion of α-lactalbumin by mammary explants taken from pregnant cows. Mammary explants from estrogen-primed or pregnant mice also require insulin, cortisol, and Prl for the accumulation of casein as well as α-lactalbumin. However, some caution is advised with wholesale extrapolation of results from culture experiments to the intact animal as well as uncritical extrapolation between species. For example, induction of the various milk proteins in culture does not necessarily reflect the timing of events in vivo. Neither do existing culture methods allow consistent synthesis and secretion of milk lipids. Finally, hormone concentrations employed may not accurately reflect the situation at the level of the mammary cell in the animal. Culture systems by their very nature represent relatively uncomplicated regulation compared with the intact animal.

Despite our seemingly good understanding of the process of lactogenesis, study of the mammary gland as a target for an apparent myriad of hormones and growth regulators remains a fertile subject matter for biochemists, molecular biologists, cancer specialists, and biotechnologists as well as dairy scientists.

Endocrine Regulation of Galactopoiesis

The term *galactopoiesis* was originally coined to describe the enhancement of an established lactation. With this strict definition, only exogenous GH (bST) and thyroid hormone are undisputed galactopoietic hormones in dairy animals. These responses also support the idea that these hormones are endogenously rate limiting. However, more generally, galactopoiesis can also be described as the maintenance of lactation. In this sense a larger number of hormones and growth factors are candidates as galactopoietic agents and are involved in the maintenance of milk production in dairy animals.

Continued secretion of galactopoietic hormones, growth factors, and regular milk removal are essential for regulation and maintenance of lactation following lactogenesis. The pituitary gland and its hormones are essential integrators of the endocrine regulation of milk secretion. This is dramatically shown by the loss of milk production in hypophysectomized lactating goats. However, milk yield can be fully restored to prehypophysectomy levels by the combined administration of Prl, GH, glucocorticoids, and triiodothyronine (T_3) as illustrated in Figure 7.6. Although species differences exist, endocrine organ ablation/replacement studies show that Prl, GH, glucocorticoids, and thyroid hormones are typically required for the full maintenance of lactation (Topper and Freeman, 1980). This does not, however, mean that additional hormones and growth factors (both humoral and local, identified and unidentified) might not also be important or mediate effects of these hormones.

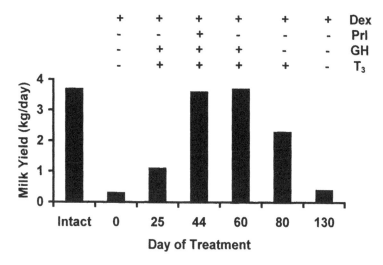

Fig. 7.6 Daily milk yields of a goat after hypophysectomy and during hormone replacement therapy. The goat was producing approximately 3700 ml of milk per day prior to hypophysectomy (intact). After treatment with dexamethasone for 2 months, milk production was as shown at time 0 in the figure. Milk production is depicted after addition or removal of the hormones. GH = somatotropin, and Dex = dexamethasone. Adapted from Cowie, 1969.

In addition to hormones that actively support synthesis of milk components, frequent emptying of the mammary gland is also critical (see "Role of Milk Removal in Milk Secretion" in Chap. 3). This process is supported by the milking-induced release of the posterior pituitary hormone oxytocin. Physiological support for both processes—milk synthesis and milk removal—is necessary for maintenance of lactation. Other factors that affect the maintenance of lactation include those that maintain the alveolar secretory cell population by slowing cell loss or by increasing cell proliferation. These factors do not impact the secretory capacity of existing cells but would change secretory capacity of the mammary gland and therefore the shape and length of the lactation curve. The following discussion considers secretion of hormones at the time of milking and their possible importance as galactopoietic agents. However, particular emphasis is given to use of recombinant bovine GH (bST) as a tool to stimulate milk production in lactating cows.

Hormone Secretion with Milking and Suckling

As discussed in earlier sections, secretion of oxytocin at milking or suckling is important in most species for efficient removal of accumulated milk and in some species is essential to obtain any milk at all. The importance of regular milk removal to prevent mammary involution has been known for many years, but it has also long been hypothesized that the secretion of galactopoietic hormones with milking or sucking was also important. Secretion of Prl and oxytocin as well as related secretions of norepinephrine or epinephrine, which can especially impact the secretion of oxytocin, are most often associated with hormone secretion at milking and effects on milk synthesis. However, glucocorticoids are also secreted in response to milking or suckling, as is GH in some species. The following sections discuss effects of several galactopoietic hormones (Akers, 1985; Capuco and Akers, 2001; Tucker, 1994; Tucker, 2000).

Hormones and Galactopoiesis

Prolactin

It is has been demonstrated in a variety of species (cows, goats, sheep, humans, and rats) that stimuli associated with milking or suckling promote the secretion of Prl. However, the secretion of Prl or other hormones in response to milking or suckling does not directly depend on the removal of secretions, since it is known that teat stimulation alone can also induce Prl secretion in nonlactating animals. Little is known about the development of this neuroendocrine reflex, but it is affected by stage of development. For example, in an experiment with 3-, 6-, and 10-month-old heifers, three of six 3-month-old heifers showed a moderate increase in blood Prl with mimic hand milking, but only two of six for the 6-month-old heifers and none of six for the heifers at 10 months of age. In contrast, for heifers tested during gestation, all of the heifers responded, and Prl secretion increased monotonically as gestation advanced. The response continues into lactation with Prl routinely secreted with each milking (Fig. 7.7), but in cows the magnitude of the response declines with advancing lactation (Akers and Lefcourt, 1982).

It is clear that the mammary epithelium responds directly to Prl. Isolated mammary epithelial cells or mammary tissues increase synthesis and secretion of mammary-specific milk proteins in response to the addition of Prl (see Figs. 7.1 and 7.2). Receptors for the hormone have been identified in mammary epithelial cells of numerous species and are classified as members of the cytokine receptor superfamily. The receptors do not have intrinsic kinase activity but associate with the Janus kinase (JAK) family of tyrosine kinases that, when activated by Prl receptor binding, dimerize and phosphorylate cellular proteins. Phosphorylation of the signal transducer and activator of transcription-5a (Stat5a) appears especially critical. For example, Stat5a is a key regulator of casein transcription in keeping with a role for Prl in lactogenesis and galactopoiesis (Hynes et al., 1997; Rosen et al., 1999).

The requirement for Prl maintenance of lactation varies with species. For example, it was shown many years ago that hypophysectomy inhibits lactation in rats and mice. As shown in Figure 7.6, Prl is also important in restoration of milk

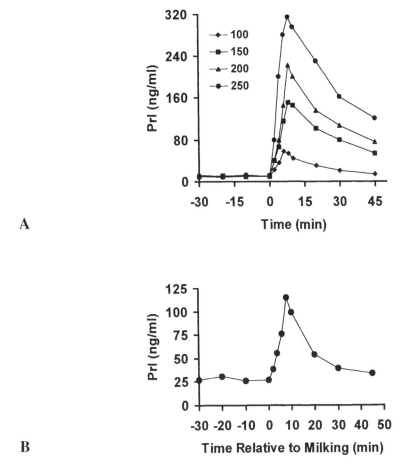

A

B

Fig. 7.7 **A** shows changes in serum Prl concentration before, during, and after teat stimulation (mimic hand milking) of pregnant Holstein heifers sampled at 100, 150, 200, or 250 days of gestation. Note the monotonic increase in response with advancing gestation. **B** shows Prl secretion in heifers in response to machine milking 30 days postpartum. Data adapted from Akers and Lefcourt, 1982.

production in hypophysectomized goats. Similar studies have not been done in cattle, but experiments with drugs to reduce circulating Prl in cows, once lactation is established, have generally shown that markedly reducing the circulating concentrations of prolactin has little apparent effect on milk production. It is also know that increasing Prl via treatment of cows with exogenous Prl has little effect on milk production or composition (Plaut et al., 1987). This suggests that circulating concentrations are not rate limiting to continuing milk secretion in cows. This contrasts with the period of lactogenesis, since Prl secretion at this time is essential to the onset and initial establishment of lactation. It is important to appreciate that even suppression of Prl secretion with CB154 or other drugs during lactation does not completely prevent secretion of Prl. Perhaps only low concentrations of Prl are able to satisfy requirements of the bovine mammary gland. However, these observations support the concept that Prl plays a key role in lactogenesis in ruminants but certainly a lesser role in galactopoiesis. In ruminants, Prl is a primary lactogenic hormone, but its galactopoietic activity is subtle and its mammogenic activity questionable. In these animals, GH is thought to be the primary galactopoietic hormone and Prl a minor player. The opposite appears to be the case in other species (e.g., rats).

In contrast with this, exogenous Prl is very clearly galactopoietic in many other lactating animals. For example, milk production is abruptly increased in rabbits given Prl. In rats Prl treatment in early lactation stimulates milk secretion and reduces the time needed for the mammary gland to refill after a period of suckling. There is evidence that Prl treatment can induce the liver to secrete a factor that can stimulate casein secretion in cultured mammary cells. It has also been suggested that mammary cell uptake and/or processing of Prl released at milking or suckling might be especially important in stimulation of mammary cell metabolism. For example, the greatest uptake of Prl by the bovine mammary gland coincides with the acute secretion at the time of milking. Some of this Prl is transported into the secreted milk. In rats and cows the suckling or milking induced release of Prl changes in correspondence with milk yield. Specifically, concentrations of Prl measured in blood taken 5 minutes after milking are higher in early than late lactation in conjunction with the decline in milk production. Concentration of Prl receptor in mammary tissue of rodents also correlates well with milk yields. The number of receptors increases to the time of peak milk production and then declines. In fact receptor number is positively correlated ($r = 0.7$) with litter weight gain in mice. Although the data are limited, mammary Prl receptor in the bovine mammary gland increases dramatically in conjunction with lactogenesis and is maintained at a high concentration during the course of lactation. The metabolic clearance and secretion rate of Prl is also positively correlated with milk production across stages of lactation in cows.

In cattle, increasing the duration of daily photoperiod from 8 hours light, 16 hours dark to 16 hours light, and 8 hours dark increases Prl concentrations in the blood several-fold. And this is associated with increased milk production.

Recently, it has been shown that increased photoperiod also increases plasma IGF-I (without increasing GH), and this has been hypothesized to mediate the milk production effect of photoperiod. Still, the ability of IGF-I to induce an increase in milk production is not certain, and subtle effects of Prl on milk production, particularly long-term effects, are possible. Regardless of Prl's ability or lack of ability to increase milk production in a relatively short time frame, recent data suggest that it may help to maintain the population of mammary secretory cells and thus promote lactational persistency. This was first proposed by Flint and Gardner (1994), who discovered that treatment of lactating rats with ergot alkaloids decreased milk production approximately 50 percent, compromised epithelial tight-junction integrity, and reduced DNA content of the mammary gland by 20–25 percent. When Prl concentrations were reduced significantly, increased epithelial cell apoptosis occurred, with an accompanying disruption of the blood-milk barrier and decline in number of secretory cells. Maintenance of the mammary epithelial cell population appears to involve an interaction between the Prl and GH axes. However, this requires further investigation to fully extend these findings to ruminants (Flint et al., 2000).

Glucocorticoids

In rats and mice it has been known that adrenalectomy severely reduces milk yield and conversely that administration of glucocorticoids to intact animals increases milk yield by retarding the decline that occurs with advancing lactation. Because the decline in milk yield during a murine lactation is believed to be due to decreased activity of mammary secretory cells, rather than a decrease in cell number, it follows that the galactopoietic effect of glucocorticoids in rodents occurs because they are rate limiting to milk synthesis.

The primary glucocorticoid in cows is cortisol. It has a major role in differentiation of the mammary alveolar secretory cells in the final stages of lactogenesis and promotion of transcription of the caseins and α-lactalbumin genes. For example, experiments in the 1970s demonstrated that additions of cortisol to mammary explants from mice midway through gestation or from estrogen-primed virgin mice promoted a dramatic increase in the appearance of rough endoplasmic reticulum and, along with Prl, the appearance of Golgi and secretory vesicles. These effects have been largely confirmed in experiments with ruminant mammary tissue. It is also known that injections of glucocorticoids into nonlactating cows with developed lobulo-alveolar tissue can induce the onset of lactation but that yields are increased with treatments that also promote secretion of Prl. These effects provide additional support for the widely held view that full lactogenesis depends on a complex of hormones acting synergistically to maximal effect. In general the glucocorticoid concentration in blood is low, and with the exception of stress-induced secretion episodes, it is rather stable through most of gestation. The increase, which occurs at the time of parturition, is most likely stress induced, and it occurs too late to be critical for at least the initial stages of lactogenesis. This does not, however, mean that glucocorticoids are not important for lactogenesis, only that basal concentrations are likely sufficient. As would be expected, specific receptors for glucocorticoids are present in mammary cells. The number per cell increases during the course of gestation and is about three-fold greater during the last trimester of pregnancy.

Milking or suckling stimulates the secretion of adrenocorticotropic hormone, which then promotes secretion of glucocorticoids. In lactating rats secretion of glucocorticoids at the time of nursing declines with advancing lactation, but in cows the response to milking persists largely unchanged throughout lactation. Overall, concentrations of cortisol are approximately doubled in lactating compared with nonlactating cows. Since adrenalectomy reduces milk synthesis, it is clear that adrenal secretions are necessary for full lactation, but it is nonetheless difficult to distinguish effects on homeostasis from effects specific to the mammary gland. Certainly lactating, adrenalectomized animals given both glucocorticoids and mineralocorticoids are more responsive than those given only glucocorticoid replacement therapy.

An interesting aspect of glucocorticoid secretion is that glucocorticoids in circulation are bound to a protein, glucocorticoid-binding globulin (CBG) that effectively inactivates the hormone since the concentration of free hormone available for binding to mammary cell receptors is minimized. During the periparturient period CBG in blood decreases. This means that the biologically effective concentration of glucocorticoid is increased, even without a change in total cortisol concentration in the blood. Mammary tissue uptake of cortisol is about two-fold greater in lactating compared with nonlactating cows, and uptake occurs most readily at the time of milking or suckling. Concentrations of intracellular free cortisol are increased in lactating mammary tissue. Along with the presence of increased receptors, these data suggest a role of glucocorticoids during lactation.

Somewhat paradoxically, most studies show that exogenous administration of adrenocorticotropin or glucocorticoids inhibits milk production. This may be a function of the treatment dose since high concentrations may impair the milk ejection reflex. Therapeutic doses of dexamethasone or other glucocorticoid derivatives given to cows suppress milk yields, but lower doses or single treatments have been reported to increase milk production. Conflicting effects of glucocorticoid administration to cows may reflect nonmammary-related responses or pharmacological effects unrelated to normal mammary function. In summary it seems likely that glucocorticoids are especially important for differentiation of the alveolar cells at the time of lactogenesis and that at least a minimal concentration is important for maintenance of lactation either because of direct mammary effects or via support of homeostasis.

It was determined that glucocorticoids bind to specific glucocorticoid receptors in mammary tissue and regulate the secretion of α-lactalbumin and β-casein. Some of these actions are synergistic with other regulatory hormones such as Prl. For ruminants there is little evidence that the glucocorticoids are limiting to milk production. However, adrenalectomy reduces milk yield that is restored by glucocorticoid treatment. Bovine mammary tissue contains glucocorticoid receptors that are present in greater concentration in lactating than in prepartum tissues, and receptor number correlates with glucose uptake. Certainly, glucocorticoids are important for maintenance of milk production in dairy species.

Oxytocin

Although an absolute need for oxytocin secretion at the time of milking or suckling varies with species, supplemental oxytocin given immediately before milking increases milk production throughout lactation in cows. As illustrated in Figure 7.8, this increase in yield parallels the normal lactation curve. This suggests that

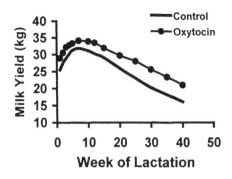

Fig. 7.8 Lactation curves of cows treated with control or oxytocin at the time of milking. Data adapted from Nostrand et al., 1991.

the response is most likely a function of improved milk removal with each milking rather than an effect of oxytocin directly on alveolar secretory cell function. In this regard, the response is similar to the pattern of increase in milk yield measured when cows are milked three times per day compared with twice daily.

Insulin

Insulin is clearly a hormone that plays an important role in the regulation of nutrient utilization during lactation. In ruminant dairy animals, insulin has no effect on the mammary uptake of glucose, acetate, β-hydroxybutyrate, and amino acids, but exogenous insulin inhibits milk production by virtue of its metabolic effects on other tissues. For example, in adipose tissue insulin promotes the uptake of glucose and acetate and stimulates lipogenesis while inhibiting lipolysis, and in liver it inhibits gluconeogenesis. It is unlikely that insulin plays a role in regulating the number of mammary epithelial cells. Mammogenic properties that were historically attributed to insulin on the basis of its effects on mammary cells in vitro can be dismissed as an artifact. Due to the use of supraphysiological concentrations of insulin in these systems, the mitogenic activity observed is attributed to the ability of insulin at high concentration to cross-react with IGF receptors and elicit IGF-related responses.

Thyroid Hormones

It has been known that the thyroid is involved in maintenance of lactation since reports in the early 1900s showed that milk yield was reduced in thyroidectomized goats. By the 1930s it was shown that thyroidectomy of dairy cows reduced milk yield and conversely that treatment with the thyroid hormone, thyroxine (T_4), increased milk yield by approximately 20 percent. Because thyroxine is efficacious when fed, these reports stimulated much interest in the practical utilization of the hormone to increase milk production in cattle. This became economically feasible by the relatively low-cost manufacture of thyroxine and thyroactive-iodinated proteins. However, results of multiple studies showed that while feeding thyroxine (or iodinated proteins) milk production increased by 10–40 percent, the galactopoietic effect was of variable duration, and milk production returned to normal or below normal levels despite continued treatment. The galactopoietic effect of thyroxine supplementation depends on a general increase in body metabolism. Thus it is not effective when cows are in early lactation (negative energy balance) and already mobilizing body reserves to meet the energy demands of lactation. A general increase in body metabolism at this time would be counterproductive to meeting the nutrient demands of lactation. It was concluded that thyroxine treatment should not be initiated before mid-lactation and that the energy density of the diet should be increased during treatment because feed intake does not increase in proportion to increased energy utilization. Furthermore, upon withdrawal of treatment, a hypothyroid condition ensues that exacerbates the usual decline in milk yield in late lactation. Despite the initial interest in thyroid hormone supplementation to increase milk yield, the temporary nature of the milk yield response and frequent undershoot below normal production led to the conclusion that its adaptation would be of minimal value.

In addition to a general affect on metabolic rate, thyroid hormones also synergize to increase the effectiveness of other lactogenic and galactopoietic hormones. For example, triiodothyronine (T_3) enhances the ability of Prl to stimulate lactose synthetase in cultured mouse mammary tissue approximately five-fold, and enhances Prl stimulation of casein synthesis in rabbit mammary tissue culture. Similarly, for estradiol to stimulate lactogenesis in mouse or bovine mammary tissue culture, T_3 must be present in the culture medium. Thus, in contrast to the general increase in metabolism evident with thyroid hormone supplementation, organ-specific changes in thyroid hormone metabolism during lactogenesis likely facilitate adaptation to a lactational state by promoting differential rates of energy utilization.

Although T_4 is the predominant thyroid hormone in the circulation, it essentially serves as a prohormone because it has little if any biological activity. The most metabolically active thyroid hormone, 3,3', 5-triiodothyronine, is produced by enzymatic 5'deiodination of T_4 within the thyroid and peripheral tissues. Extrathyroidal activity of thyroxine-5'-deiodinase (5'D) is an important regulator for altering localized T_3 availability in many tissues. Activity of the enzyme also varies with physiological state. With onset of lactation in rodents and ruminants, there is an increase in 5'D in the mammary gland and a decrease in the liver. These changes are believed to maintain a euthyroid state in the lactating mammary gland despite the fact that the body is hypothyroid as a whole. The transfer of iodine, iodinated nonhormonal compounds, and thyroid hormones through the mammary gland into the milk further exacerbates a systemic hypothyroid condition. Maintenance of a euthyroid state in the lactating mammary gland in the midst of a functional hypothyroid condition is consistent with increasing the metabolic priority of the mammary gland and providing T_3 to heighten the effect of other galactopoietic hormones. For example, this occurs in response to treatment of cows with exogenous bST (Capuco et al., 1989).

Changes in thyroid hormone metabolism are proportional to lactational intensity in rodents and are involved in eliciting a response to galactopoietic hormones. In mice, thyroid hormones are necessary to obtain milk production increases in response to treatment with GH or Prl, and mammary 5'D is specifically responsive to galactopoietic hormones, since GH and Prl increase mammary 5'D activity but do not alter activity in liver or kidney. In parallel, changes in 5'D activity are hypothesized to at least partially mediate or augment the galactopoietic effect of bST in dairy cows. However, results are not as consistent as those in rodents, and further investigation is necessary to clarify the interaction between thyroid hormone metabolism and bST-increased milk production in cattle.

The relationship between GH and thyroid hormones is not limited to GH-induced alterations in 5'D during lactation and galactopoiesis. There is a close relationship between thyroid hormones, thyroid hormone metabolism, and GH and IGF-I synthesis. Mechanistically, T_3 can alter hepatic GH receptor binding and thus enhance GH stimulation of IGF-I synthesis. Alternatively, T_3 can also increase IGF-I synthesis in the absence of GH. It is worth noting that in those situations when GH does not stimulate IGF synthesis (e.g., during food restriction, fetal development, sex-linked dwarfism, and hypothyroidism), there is evidence for T_3 deficiency. In addition, T_3 serves as a regulator of GH synthesis by the pituitary. Conversely, GH can alter synthesis of 5'D and therefore peripheral production of T_3 (Capuco et al., 1989).

Somatotropin

As outlined in recent reviews (Etherton and Bauman, 1998; Bauman, 1999), by the 1920s it was discovered that crude extracts prepared from homogenates of bovine pituitary glands could stimulate the growth of rats. The active agent was dubbed *growth hormone* or *somatotropin* after the Greek word for growth. Soon thereafter, the ability of such extracts to promote milk secretion in pseudopregnant rabbits and milk production in lactating goats was reported. Some of the more extensive early experiments with cows by the Russians Asimov and Krouze in the 1930s involved the treatment of more than 2000 cows with crude anterior pituitary extracts. They consistently observed that injections of the extract produced a short-lived increase in milk yield. They further noted the "absolute harmlessness" of the injections and that responses were "more profitable on a well-run farm than a farm with a poor food basis or where cattle were kept under unsatisfactory conditions." Soon after, as part of efforts to increase food production during World War II, the British scientists Folley and Young and colleagues also studied the effects of GH on milk production in cows and goats. They identified GH as the primary active galactopoietic component in bovine pituitary extracts. Other studies established dose response curves and confirmed that relative responses were greater in declining lactation, that gross milk composition was unaffected, and the generally potent effect of GH on milk yield in dairy ruminants. They concluded that use of pituitary GH would be highly profitable to individual farmers but that an inadequate supply limited the impact GH could have in stimulating the national milk supply. For example, approximately 25 pituitaries are needed to produce enough GH for a typical daily treatment.

A common belief related to nutrient regulation at the time of these studies was that nutrient use largely involved a competition between organs for needed substrates. This means that an increase in milk production in response to, for example, treatment with GH would require either an increase in the metabolism of the mammary gland to better "fight" for nutrients or a reduction in nutrient use by other tissues to support increased nutrient demands of the mammary cells for milk synthesis and secretion. Mechanisms for GH's actions were generally thought to depend on acute effects to "push" metabolism of peripheral tissue to reduce competition and thus favor the lactating mammary gland. On the surface many acute effects of GH seem to support this model. These GH-stimulated responses include glycotropic (reduced response to insulin in glucose tolerance testing), diabetogenic (hyperglycemia and glycouremia), and lipolytic (increases in blood nonesterified fatty acids) activity. Given these effects, a number of researchers were concerned that GH induction of these responses would promote metabolic problems in lactating cows. Many of the initial trials with pituitary-derived GH were with cows with lower milk yields generally or with cows after peak lactation, and treatments were often only administered for a few days. However, some of these studies lasted 10–12 weeks with milk yield responses of up to 40 percent without apparent metabolic problems.

By the late 1970s researchers at the National Institute for Research in Dairying in the United Kingdom and at Cornell University began to question more closely this dogma that prolonged treatment with GH would necessarily produce adverse metabolic effects in lactating dairy cows. Three related events fueled this re-examination. One was growing support for the idea that genetically superior dairy cows secreted higher concentrations of GH in their blood (Akers, 2000). One of these initial studies showed that Friesian cows had greater circulating concentrations of GH than Hereford × Friesian cows during lactation. Concerns were raised over differences being mediated by energy balance rather than genetic merit. However, New Zealand researchers subsequently reported greater GH in high genetic merit than low genetic merit cows. Moreover, in a much larger study, basal GH concentrations were consistently greater in Holstein cows selected for increased yield compared with controls during each of three stages of lactation, in first or second lactation, and if milking twice or three times daily. These researchers also found greater GH response to thyrotropin-releasing hormone in selection line cows than in controls. Differences existed in net energy balance in twice versus three-times daily milked cows and at each of three stages of lactation (30, 90, and 200 days in milk). However, overall there was no significant difference in energy balance between selection and control line cows. Therefore, differences in GH secretion were related to genetic selection and not confounded by differences in energy balance. An example is illustrated in Figure 7.9. Overall, these results support the idea that GH is at least a partial mediator of selection pressure for increased milk yield (Barnes et al., 1985; Kazmer et al., 1986).

The second development that led scientists to question the belief that prolonged treatment with GH would necessarily produce adverse metabolic effects in lactating dairy cows was expansion of the concept of homeorrhesis to apply to the high production of genetically superior cows. The overriding idea was that homeorrhesis does not act to defeat homeostasis but rather that chronic changes in physiological controls allow coordination of physiological processes to support major

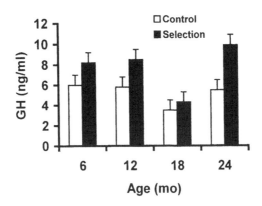

A

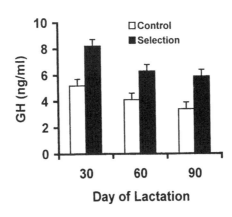

B

Fig. 7.9 Basal GH secretion in genetically selected Holstein cattle is illustrated. **A** shows data for heifers and **B** data for heifers during the first lactation. Concentrations of GH are consistently elevated in genetically superior (selection line) animals. Data adapted from Barnes et al., 1985, and Kazmer et al., 1986.

physiological events (e.g., fetal development, reproductive performance, or in this case a sustained high level of milk production) while homeostasis is preserved. Initial advances toward understanding of GH as a homeorrhetic agent were achieved using pituitary-derived somatotropin.

The clearest galactopoietic activity of GH appears to be in coordinating changes in tissue metabolism that promote a flux of nutrients and energy to the mammary gland. Adipose and liver tissues have especially prominent roles in this process, and the effects appear to be largely mediated by GH receptors in these tissues. In adipose tissue, GH inhibits lipogenesis when animals are in positive energy balance and promotes lipolysis when animals are in negative energy balance. In liver, GH promotes gluconeogenesis, critical in ruminants where nutrient absorption directly provides only a negligible percentage of the glucose required for milk synthesis. In other tissues, the galactopoietic effects of GH are believed to be mediated by other members of the somatotropin axis (i.e., IGF-I and IGFBPs). In skeletal muscle especially, GH decreases glucose utilization and oxidation of amino acids. This serves to conserve nutrients and energy for synthesis of milk lactose, protein, and lipid. A direct action of GH has not been observed in lactating mammary tissue, and functional ligand-binding assays have failed to detect receptor protein. This supports the idea that the IGFs and their binding proteins are probable mediators of GH effects on the lactating mammary gland.

The third development that led to changing beliefs about the effects of prolonged treatment with GH was the evolution of recombinant DNA technology to allow production of large quantities of pure bovine GH. With respect to cattle Genetech, Inc., largely pioneered these efforts in combination with Monsanto Company to commercialize agricultural use of recombinant GH (bST) in animal agriculture. As with other currently marketed recombinant proteins (e.g., human GH, human insulin, and interferon), the essential technology involved the splicing of the cDNA coding for bovine GH into the plasmid DNA of a laboratory strain of *Escherichia coli*. Once incorporated into the microorganism, the bacteria are grown in fermentation vats, where the cells produce bST along with their own cellular proteins. The cells are subsequently killed, the cells ruptured, and the accumulated bST is purified. Bovine pituitary cells actually synthesize four isoforms of GH believed to result from alternative splicing or processing of the GH mRNA. These proteins vary slightly from one another by one or two amino acids of the 190 or 191 in the native molecule. The recombinant bovine GH is identical in sequence with the first 190 amino acids of one of these isoforms but does have the amino acid methionine substituted for the usual amino acid alanine at the amino (NH_2) terminal of the protein. The development of this technology and availability of relatively large quantities of bST ushered in an amazing period of intense research activity on the effects of bST on milk production and animal health in both lactating and nonlactating dairy cows and heifers. To be sure, the companies involved were interested in eventual commercialization, but the wealth of research on bST and mammary function was unprecedented and voluminous.

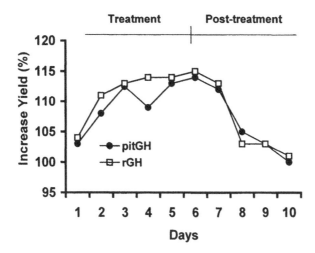

Fig. 7.10   Comparison of the effect of pituitary-derived versus recombinant bovine GH on milk production in lactating cows. Cows were given daily injections of 25 mg of GH for 6 days. Pretreatment milk production averaged 32 kg/day. Data adapted from Bauman, 1999.

Figure 7.10 illustrates a comparison of milk production in cows treated with pituitary-derived compared with recombinant bovine GH.

With the production of recombinant DNA–derived bST, it became feasible to utilize the hormone for increasing lactational performance of dairy cows. Subsequent investigations and now commercial use have rapidly expanded knowledge of the physiological effects of bST and demonstrated its efficacy and certainly its overall safety as a stimulant of milk production. The efficacy of bST to increase milk production has also been demonstrated for goats, sheep, and buffaloes. Indeed, somatotropin appears to be the primary galactopoietic hormone (i.e., hormone that increases milk production) in mammals, except for rodents, where Prl appears to be the primary galactopoietic hormone.

Somatotropin Effects on Milk Production

Administration of bST to lactating dairy cows increases the yield and efficiency of milk production. In response to injection of bST, milk secretion increases within a day and is maximized within a week. The increased milk yield is maintained as long as treatment is continued but quickly returns to control levels when bST is discontinued. The milk yield response is dose dependent and the response curve is hyperbolic in shape. At approximately 40 mg of bST/day nearly maximal response is obtained. Milk yield achieved with near maximal doses of bST is impressive, with increases reported as high as 30–40 percent. Typically, bST increases milk production by 4–6 kg/day, approximately a 10–15 percent increase in yield. The magnitude of response to a particular dose of bST depends upon biological variation, stage of lactation, and management parameters.

During midlactation the pattern of bST administration does not affect the magnitude of the galactopoietic response. Similar increases in milk yield are obtained with the same daily dose of bST whether administered as once daily injections, 4-hour pulses, or a constant infusion. The bST formulation currently approved for use in the United States is a prolonged-release n-methionyl-bST (Posilac, Monsanto Co.) that was approved by the U.S. Food and Drug Administration in November 1993. It is administered at a dose of 500 mg per cow every 2 weeks. Package instructions are that treatment should be initiated after peak lactation at >60 days postcalving, when cows are at or near positive energy balance. During early lactation, response to bST is minimal. Figure 7.11 shows milk yield responses to bST in dairy herds in the northeastern United States. From Dairy Herd Improvement records and sales data for bST, two groups of herds were identified, those in which bST was used continuously in at least 50 percent of the cows and control herds that were bST-free. Data are given for an 8-year period, which includes 4 years prior to FDA approval and 4 years after FDA approval of bST usage. The data are from 340 herds and more than 80,000 cows and 200,000 lactations. The first year after approval average milk yield increased 3 kg, and this level of increase for bST-adopting herds has been maintained. Since not all cows in the bST herds are treated (i.e., some cows are not eligible < the 60 days postpartum), this is a conservative estimate of the effect of bST treatment on milk production (Bauman et al., 1999).

In addition to increasing milk yield, bST increases the efficiency of lactation. Treated cows increase feed intake over the first few weeks to largely match

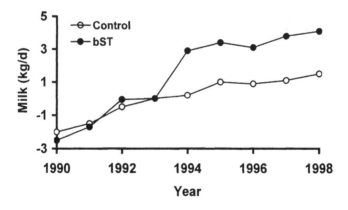

Fig. 7.11 Milk production in northeastern U.S. dairy herds before and after approval of bST. Use of bST began in February 1994. Data cover a period before and after approval; bST was never purchased for control herds, and bST was used with bST herds during the entire period after approval. Data represent 340 herds, 80,000 cows, and 200,000 lactations. Yields are expressed relative to 1993, the year prior to approval. Adapted from Bauman et al., 1999.

increased nutrient demands for milk synthesis so that cows typically remain in a neutral or positive energy balance during the majority of lactation. However, because milk secretion increases more rapidly than voluntary intake at the start of bST treatment, these cows initially experience a temporary period of negative energy balance. Respiration calorimetry studies demonstrated that the energy requirements for body maintenance and the partial efficiency of milk synthesis from absorbed nutrients were not changed in bST-treated cows. This means that the energetic efficiency of milk production is increased in bST-treated cows because the increased milk production is achieved without nutritional overhead. For example, with an 11 percent increase in milk production, nine bST-treated cows can yield the same amount of milk as 10 control cows, and the energy savings will equal the maintenance requirements for one cow. However, it is important to note that bST is not unique in this regard. There are other methods of increasing milk production, such as increased milking frequency, that improve efficiency of lactation because production is enhanced without increasing energy requirements for maintenance. In contrast, thyroid hormone supplementation increases milk production but also increases body metabolism and maintenance requirements so that there is no gain in efficiency.

Administration of bST typically does not alter the gross composition of milk from cows in positive energy balance. Synthesis of milk proteins, fat, and lactose are all increased proportionately to milk volume so that normal milk composition is maintained. Additionally, change in the composition of milk protein or milk lipid is minimal with bST treatment. Casein proteins are expressed in the same proportions in milk from control and bST-treated cows, whey proteins that have been evaluated appear similarly unaffected by treatment, and the ratio of whey to casein proteins is unaltered. Lipid classes and fatty acid composition of milk fat are not altered or altered very slightly by bST treatment when cows are in positive or neutral energy balance. For example, there may be a small increase in the relative amount of long-chain fatty acids in the milk of bST-treated cows. For reasons discussed subsequently, the fat content of milk increases, and fatty acid composition may change, if cows are in negative energy balance when bST is administered. Mineral content of milk appears largely unaffected by bST treatment, and vitamin concentration is seemingly unaltered, although vitamin content has been less thoroughly examined.

When bST is provided as a sustained release formulation, small cyclical effects on milk yield and composition have been noted. With the biweekly injection protocol, milk production peaks 7–9 days after injection and then declines until the next injection, seemingly as a function of bST concentrations in the blood. Concentration of milk lactose follows the same cyclical pattern as milk yield, although the reasons for this effect are unclear. Milk fat and protein cycle in a manner that is out of phase with milk yield (i.e., the concentration of milk fat and protein is at a nadir when milk yield peaks). With the biweekly injection protocol, a steady state of the metabolic alterations coordinated by bST is seemingly never fully achieved. Thus, synthetic processes for synthesis of milk components may not be fully coordinated, resulting in minor and temporary alterations in milk composition. Other factors, such as changes in nutrient balance or changes in mammary blood flow, may also occur in response to biweekly injections of bST. These may produce small changes in availability of nutrients to the mammary glands and may partially underlie the small fluctuations in milk volume and composition. However, it should be noted that these cyclical fluctuations in composition are not apparent in the bulk tank milk because cows within a herd typically calve asynchronously and are injected with bST asynchronously. Thus, this effect is of no importance to milk processors or consumers. Indeed, these variations in milk composition during bST treatment are minor compared with normal variation in milk composition that occurs between herds and within a herd. Milk composition is more strongly influenced by season, stage of lactation, genetics, nutritional management, and energy balance.

Mode of Action for bST

The galactopoietic action of exogenous bST probably comes from a combination of direct and indirect effects: (1) direct stimulation of mammary tissue, (2) indirect stimulation of mammary tissue, (3) direct effects on other tissues to supply nutrients to support increased milk production, and (4) indirect effects on other tissues to supply nutrients to support milk production. The preponderance of evidence suggests that bST enhances milk production largely by partitioning nutrients to support milk production, by both direct and indirect actions of bST, but bST does not alter digestibility of nutrients. Effects on the mammary gland appear to be indirect, whether direct effects of bST are operative remain to be determined.

Lipid metabolism in adipocytes is chronically affected by bST administration. When cows are in negative energy balance during bST administration, lipolysis is increased. This is noted by increases in blood nonesterified fatty acids, increased milk fat percentage, and an increase in the percentage of long-chain fatty acids in the milk fat. (In the mammary gland, the long-chain fatty acids incorporated into milk triacylglycerols derive from mobilized fat stores and from dietary sources, whereas short- and medium-chain fatty acids are synthesized within mammary tissue.) When cows are in positive energy balance during bST administration, lipogenesis is inhibited. These changes depend on altered responsiveness to key homeostatic signals and corresponding changes in the activities of key enzymes.

Adjustments in insulin responsiveness serve to regulate lipid metabolism, and bST antagonizes some of the actions of insulin. Insulin promotes the facilitated transport of glucose into most cells of the body (the central nervous system and mammary gland are not insulin dependent) and inhibits many of the liver enzymes that catalyze gluconeogenesis. It also promotes the synthesis of glycogen and inhibits glycogenolysis. It stimulates the deposition of fat by enhancing the activity of key enzymes in fatty acid synthesis and by inhibiting lipolysis of triacylglycerol. It stimulates protein deposition by enhancing facilitated uptake of amino acids and increasing the activity of some ribosomal enzymes involved in protein synthesis. Only a few of insulin's actions are antagonized by bST treatment. Most importantly, bST inhibits the lipogenic activity of insulin, and the effect appears to be exerted on processes that are downstream in the signaling cascade from the insulin receptor. This is consistent with the idea that bST inhibits a limited number of insulin actions. In addition, lipoprotein lipase (LpL) is an enzyme that is partly regulated by insulin and is decreased by bST treatment of lactating dairy cows. Specifically, there is a decrease in LpL in adipose tissue but no change in mammary tissue. LpL hydrolyzes triacylglycerols of very low density lipoproteins and chylomicrons in the serum, permitting the uptake of the nonesterified fatty acids by surrounding cells. Reduced LpL in adipose tissue and maintenance of normal LpL in mammary tissue, along with the inhibition of lipogenesis in adipose tissue, promotes the preferential delivery of nonesterified fatty acids to the mammary gland for synthesis of milk fat. Bovine somatotropin also decreases adipose tissue expression of fatty acid synthase and acetyl-CoA carboxylase, key enzymes involved in lipogenesis.

When nutrients are in limited supply, bST conversely stimulates lipolysis again by altering the response to homeostatic regulators. Dairy cows treated with bST mobilize considerably more nonesterified fatty acids following epinephrine challenge than do control cows. However, there is little change in adrenergic receptor numbers and no change in the stimulatory G proteins and other components of the cyclic-AMP lipolytic signaling pathway. Rather, it was shown that bST enhances lipolysis by antagonizing antilipolytic regulators, particularly by a decreased activity of the inhibitory G proteins. Thus, bST promotes lipolysis by chronic inhibition of antilipolytic regulation. Enhanced lipolysis often comes into play when bST treatment is initiated. Because cows are typically near neutral energy balance when treatment is initiated and feed intake does not increase immediately, bST induces a period of negative energy balance that requires the mobilization of energy stores. When cows subsequently enter positive energy balance, for the majority of lactation, inhibition of lipogenesis becomes the hallmark of bST action on lipid metabolism.

Carbohydrate metabolism is also altered by bST treatment to meet the increased glucose requirement for greater milk secretion. Increasing hepatic glucose production and decreasing oxidation by body tissues derive increased glucose for milk synthesis. In ruminants, the products of rumen fermentation are the volatile fatty acids, and only a small percentage (15 percent) of blood glucose is derived from the diet. Body glucose supply is met by hepatic gluconeogenesis, which can amount to production of 3 kg/day in a lactating cow. Enhanced hepatic gluconeogenesis in bST-treated cows is at least partly explained by antagonizing insulin's ability to inhibit gluconeogenesis. Glucose serves as the substrate for lactose synthesis in the mammary gland, and in high-producing lactating dairy cows nearly 85 percent of total glucose turnover is used for milk synthesis. Treatment with bST increases net utilization of glucose in mammary tissue and decreases glucose utilization by nonmammary tissues. These metabolic adaptations are sufficient to provide the necessary glucose for milk synthesis; no glucose deficit is encountered and ketosis is not induced. Somatotropin treatment also decreases expression of glucose transporters in skeletal muscle and in fat stores but has no effect on these transporters in the mammary gland, so there is an increasing proportional flux of glucose into the mammary gland despite the lack of direct effect on mammary transporters. Other data suggest that effects on glucose uptake are primarily secondary to nutrient gradients created by metabolic effects on the tissues.

Protein metabolism of bST-treated lactating dairy cows is also altered to support the added amino acid requirements for increased milk protein synthesis. This appears to be largely the result of more efficient utilization of amino acids. Whole body oxidation of amino acids is reduced in bST-treated dairy cows, and there is a resulting decrease in concentrations of urea and decreased urinary nitrogen loss. Increased milk protein synthesis is supported primarily by the increased availability of precursors to the mammary gland due to decreased oxidation of amino acids by other tissues. The coordination between tissues in homeorrhetic regulation of bST stimulation of milk synthesis is illustrated in Figure 7.12.

Bovine somatotropin exerts an overarching control, but not an overriding control, on metabolic processes that support milk production. While bST exerts homeorrhetic regulation, homeostatic regulatory processes that ensure animal health are maintained, and other homeorrhetic mechanisms such as those that support body growth or that support fetal development during pregnancy may be operative. Increases in milk yield are greater in multiparous than in primiparous cows. This is likely because the milk response to bST in younger animals is reduced by an amount that is dictated by the nutrient requirements of continued body growth. Similarly, when bST-treated cows are simultaneously lactating and pregnant, milk production declines normally during the later months of pregnancy; this likely minimizes conflict with nutrient demands for fetal growth. Use of bST does not promote increased milk production to the detriment of a young lactating animal's continued body growth or of a lactating animal's ability to support pregnancy. Normal physiological processes that serve to ensure the well-being of a lactating animal and survival of her fetus are still operative during prolonged use of bST during lactation. Indeed, use of bST over multiple lactations has proven to be safe and effective.

Although much has been learned about the nature of the metabolic alterations and tissue-specific effects of bST, the means by which the hormone signal elicits the biological response is poorly understood. Effects of bST on adipose tissue can be demonstrated in vitro, suggesting that these effects are direct and mediated by the GH receptor. Effects on the liver are also presumed to be direct because these too can be mimicked in vitro. However, this does not rule out the possibility that locally acting paracrine or autocrine factors may also be critical. In contrast, effects of bST on muscle and mammary tissue are likely mediated by the IGF system. Although the metabolic effects of bST on nonmammary tissues effectively spare nutrients to support milk synthesis, there are also apparent effects at the level of the mammary gland. Infusion of IGFs into the local arterial supply of the mammary gland stimulates milk production, and thus, it can be argued that there are IGF-mediated effects on mammary gland synthetic ability. The association between energy balance, bST, the IGF system, and milk production also supports

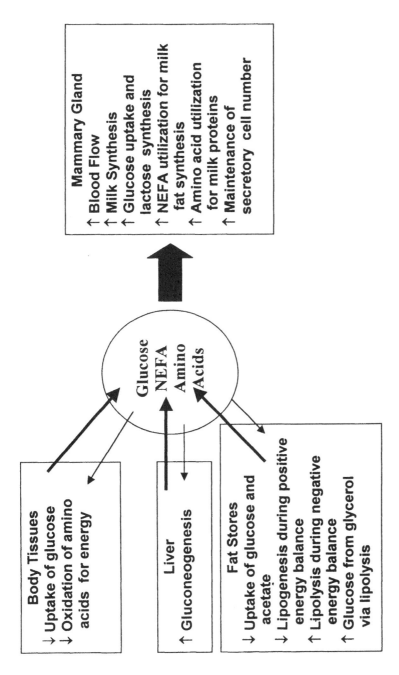

Fig. 7.12 Metabolic adaptations induced by bST. Administration of bST induces metabolic adaptations of tissues in a fashion that provides nutrient partitioning and metabolic support for increased milk production. Glucose, nonesterified fatty acids (NEFAs), and amino acids are partitioned toward the mammary gland. Data adapted from Bauman, 1999.

a role for IGF-mediated effects on the mammary gland. Moderate undernutrition causes a muted IGF response to administration of bST and a reduced galactopoietic effect. Thus, during early lactation, when cows are in negative energy balance, IGF response to bST administration is reduced, and bST is a less effective stimulator of milk production. During severe undernutrition there is dissociation between bST and the IGF system—both IGF response and the milk production response to bST are abolished. Other than bST, thyroxine is the only hormone known to increase milk production in dairy cows, and there are numerous interactions between the somatotropic and thyroid hormone axes. Indeed, tissue-specific changes in thyroid hormone metabolism alter the local action of systemic thyroid hormones, which appears important for supporting milk production and for modulating the galactopoietic response to prolactin and somatotropin in mice and rats. The situation in dairy cows is less clear and has received scant attention.

In addition to metabolic effects, recent data suggest that bST can alter cell population kinetics within the mammary gland (Capuco et al., 2001). Production data show that bST increases the persistency of lactation. This may be achieved by decreasing the loss of secretory cells during lactation and by increasing the cell proliferation. In goats bST administration increases maintenance of cell number as lactation progresses, due to decreased cell loss. Cell proliferation is increased in mammary tissue of bST-treated dairy cows and heifers during midlactation. These data are consistent with the in vitro mitogenic activity of IGF-I in bovine primary cell culture, mammary tissue slices, and an established line of bovine mammary epithelial cells (Mac-T cells) and suggest that the IGFs mediate this effect. However, decreased cell death and increased cell proliferation assist in the partial maintenance of the population of secretory cells, as previous studies have demonstrated that bST does not actually increase mammary cell number. These effects supplement the metabolic alterations induced by bST and lessen the decline in milk production with advancing lactation.

It is questionable whether bST has direct effects on the lactating mammary gland. Although bST is galactopoietic in vivo, addition of bST to mammary culture systems has failed to increase synthesis of milk components, and receptor-binding assays have failed to detect GH receptor in mammary tissue. These early results argued against a direct effect of bST on the mammary gland, and it was presumed that if endocrine stimulation of the mammary gland occurred, it was via bST-induced increases in circulating IGFs. Because bovine mammary epithelial cells have receptors for IGF-I, they are targets for IGF signaling. Indeed, infusion of IGF-I or IGF-II into the local arterial supply to one of the mammary glands of a goat caused an increase in milk production and blood flow to the infused gland within 2–4 hours. Although this increase in milk yield is consistent with a direct galactopoietic effect of IGFs on mammary tissue, it may also have been an indirect outcome of increased blood flow and nutrient supply to the mammary gland. It is interesting to note that in vitro treatment with bST increased milk fat synthesis by mammary explants when cocultured with adipose and liver explants, but not in the absence of liver and adipose tissue. This result is consistent with a nutrient and hepatic IGF-mediated effect on milk component synthesis (Keys et al., 1997).

More recently, GH receptor mRNA has been detected in lactating bovine mammary tissue by Northern blotting and by in situ hybridization. The mRNA for somatotropin receptor was localized in both epithelial and stromal (nonsecretory) elements of mammary tissue, but IGF-I mRNA was restricted to stroma. Since the somatotropin-receptor transcripts in bovine mammary tissue are translated into protein as detected by immunohistochemistry (as in the rabbit and rat), the results open the possibility that bST may also have direct effects on the mammary gland that are mediated by local production of IGFs (Sinowatz et al., 2000; Plath-Gabler et al., 2001). The importance of locally produced IGF-I is indicated by the recent demonstration of normal body growth of mice that do not produce hepatic IGF-I and have very low circulating concentrations of IGF-I (75–80 percent reduction) due to deletion of the hepatic IGF-I gene. Rather than systemic IGF-I, locally produced IGF-I may be of primary importance for (paracrine) regulation of mammary gland function (see "The Insulin-like Growth Factors Axis," Chap. 6). However, as discussed relative to mammogenesis, complexity involved in modulating IGF-regulated functions include the local concentration of IGF, expression of IGF receptors and their downstream signaling pathways, and the types and quantities of IGF-binding proteins.

Mammary epithelial cells also synthesize a number of IGFBPs. Depending upon the specific IGFBP, the binding proteins may reduce IGF activity by competing with IGF receptors for ligand, increase IGF activity by serving as delivery vehicles to the target cell, or serve as a reservoir for IGFs, causing their slow release and reducing IGF turnover. Furthermore, the IGFBPs may have activities that are independent of their interaction with IGFs, and they are subject to enzymatic modifications that may alter their various activities (Clemmons, 1998). The picture that emerges is one of a highly complex IGF system with multiple levels of regulation, making the specific actions of IGFs during lactation difficult to resolve (Cohick, 1998).

Effects on Udder Health

The effect of bST on mastitis has been studied extensively in the 1990s using more than 11,000 cows in 19 investigations. The general conclusion of these experiments is that bST treatment does not significantly alter the incidence of mastitis and has negligible effects on milk somatic cell count. In a review of Dairy Herd Improvement records from dairy herds in the northeastern United States (8 years, >80,000 cows, and >2 million test days), Bauman and coworkers reported no change in the stability and herd life of cows from herds treated with bST. Although bST caused a small but statistically significant increase in the milk somatic cell count, the increase is of little biological significance. It is consistent with the small increase in somatic cell count that accompanies increased milk yield, and the increase is considerably lower than the effects of season, parity, breed, and age in control cows. Interestingly, bST has been used as a therapeutic agent in acute mastitis. There is evidence that it enhances neutrophil function and quickens the recovery from (coliform) mastitis. It has been noted in a number of studies that the annual gain in milk yield through genetic improvement corresponds with a small but statistically significant increase in mastitis incidence. Since bST increases milk yields, it should probably not be surprising to see this trend; however, it is difficult to distinguish effects associated with increased milk production from possible effects of bST.

Given the voluminous literature and the now very widespread commercial use of bST in dairy cows in the United States and other countries, the testing and evaluation of this technology have likely made it the most intensely scrutinized new animal technology in history. Public debate about bST use was also extensive. Special interest groups and very public individuals predicted dire consequences from adoption of bST for use in dairy cows. Some of the concerns that had little if any scientific basis: bST approval would cause a massive reduction in milk consumption, milk price would decline, and farmers would go bankrupt. Others predicted dire animal consequences. Media coverage was extensive near the time of approval by the Center of Veterinary Medicine of the Food and Drug Administration in 1994, with more than 800 reports in the first quarter of the year. At the current time the regulatory agencies of more than 50 countries have reviewed safety concerns and approved bST for use. It is estimated that more 3 million cows currently receive bST supplementation. In the United States this includes animals in herds of all sizes, located in every region of the country. Scientific and anecdotal information support the view that bST is a safe, effective, and profitable management tool for the dairy farmer. Although U.S. food laws do not mandate labeling of milk from bST-treated cows because the composition of the milk from the bST cows is equivalent to that of controls, a relatively small niche market has developed for milk obtained from herds not using bST. However, since there is no chemical test to distinguish the milk from bST- and non-bST–treated cows, validation of milk from non-bST–treated cows simply depends on the producers signing a certification declaring that bST is not used on their farms. The market for milk from non-bST–treated cows is analogous to niche markets for other organic farm products. A recent estimate is that milk from these farms constitutes less that 1 percent of fluid milk sales in the United States. This suggests that the vast majority of consumers are interested in wholesome food products obtained at competitive prices. In fact during the first year after approval, fluid milk consumption in the United States increased about 1 percent.

Reasons varied for the rancorous debate on the approval of bST for use in dairy cows, but logic suggests several elements were important. This was the first proposition for widespread treatment of animals with a recombinant DNA–derived product that impacted not animal health but animal production. Second, the wholesomeness of milk and milk products seems to occupy a special position when compared with many other food materials. Third, the now roundly refuted dire predictions of consequences on animal health (Collier et al., 2001) coincided with a growing affluence and associated "greening" of attitudes among many segments of the population. Finally, popular concerns that the adoption of the technology might hasten the disappearance of family farms and change the sociology of rural farm areas were seemingly carried along on a wave of nostalgia at the time. Certainly, the debate was an interesting process to observe.

The greatest success goes to the best farm managers. Although special diets are not required for bST treatment to be beneficial, farm managers should, of course, allow animals a sufficient intake of a balanced diet to ensure increased production. An unexpected consequence of bST use includes the increased persistency of lactation perhaps making an increase in the typical 12-month calving interval desirable. On a herd basis, extended lactation length would reduce the number of births and associated problems, lower the opportunity for early postpartum metabolic problems, and likely reduce veterinary costs. Under circumstances where waste production is a problem, it would be possible to produce the same amount of milk with fewer cows. This could reduce the environmental impacts of excess manure and nutrient production. In summary, adoption of bST represented a milestone in dairy production.

Galactopoiesis and GH in Other Species

Although dairy cows have been the subject of most investigations, bST has been shown to be an effective galactopoietic hormone in other dairy animals, including sheep, goats, and the Italian water buffalo. As with dairy cows, substantial increases in milk production are obtained (14–30 percent), and the composition of milk remains unaffected. The processing attributes of milk are unimpaired. In fact, coagulation time is improved in milk from ewes during late lactation. Because sheep, goats, and buffalo are seasonal breeders, the ability to increase persistency of lactation is particularly attractive to maintain milk production throughout the year.

Insulin-like Growth Factors and Galactopoiesis

The IGF axis appears to be an essential participant in the galactopoietic response to exogenous bST. When the production of IGF-I is uncoupled from bST regulation, such as occurs during negative energy balance, then a milk production response to bST is abrogated. Infusion of IGF-I into the local arterial supply to the mammary gland of goats rapidly increased milk synthesis. These data strongly support a galactopoietic role for IGF-I. Galactopoietic activity of IGF-II is uncertain.

Additionally, IGF-I is a mammary mitogen and survival factor. The ability of IGF-I to induce cell proliferation has been demonstrated in numerous in vitro and in vivo mammary model systems. Recently, administration of bST was shown to increase the percentage of mammary epithelial cells expressing Ki-67, a nuclear antigen marker for cell proliferation, approximately three-fold (Capuco et al., 2001). It is proposed that this apparent proliferation response to bST is mediated by IGF-I. Such increased cell renewal would limit the decline in a number of mammary epithelial cells that occurs with advancing lactation and accounts for the steady decline in milk production after peak lactation. Regulation of mammary apoptosis in rats seemingly involves an interaction between Prl and IGF-I. Reduction of plasma Prl by ergot alkaloids decreases milk secretion, accompanied by an increase in the incidence of apoptosis in the mammary gland. Prl reduces the synthesis of IGF-binding protein-5 (IGFBP-5), thus limiting its ability to bind IGF-I and suppress its cell survival activity. Thus, Prl is thought to promote cell survival so that the outcome of Prl insufficiency is increased mammary apoptosis. Whether Prl regulates IGFBP-5 or an analogous IGFBP in the mammary glands of ruminants remains to be demonstrated. However, in ruminants, by increasing cell renewal or increasing cell survival, bST increases lactational persistency and maintains mammary cell number as lactation advances.

Pregnancy and Galactopoiesis

Unlike most species, dairy cows are typically pregnant through the majority of lactation, and goats in late lactation. This means that the hormones of pregnancy can impact lactation. One of these hormones is produced by the placenta (binucleate cells of the trophoblast), is a member of the somatotropin-prolactin family, and is known as placental lactogen (PL). The relative lactogenic and somatotropic activities of PL vary with species. In ruminants, PL has a greater homology with Prl than GH, but it binds to both lactogenic and somatogenic receptors. Concentrations of PL in the maternal circulation of dairy cows are very low, whereas concentrations in the maternal blood of sheep and goats are quite high (100- to 1000-fold greater than for cows). Concentrations peak during the last trimester of pregnancy. In these ruminants, exogenous PL stimulates milk production by mechanisms that differ from those that mediate the galactopoietic effects of GH. Compared with bST treatment, PL treatment of dairy cows or ewes increases milk production more slowly without increasing lipolysis. Feed intake of PL-treated cows also increases rapidly unlike the delayed increase in bST-treated cows. The biological function of PL appears related to maternal partitioning of nutrients. Under usual circumstances, the physiological impact of PL on the mammary gland is more likely on mammary growth or lactogenesis in animals that are not lactating when pregnant, and perhaps on growth and lactogenesis during the periparturient dry period in animals that are lactating during pregnancy (Byatt et al., 1992).

Despite possible galactopoietic effects of PL, concomitant pregnancy decreases the persistency of lactation, with an accelerated decline in milk production occurring during the last months of pregnancy in cows. The reasons for these declines are unclear, but physiological processes that promote fetal growth likely take precedence over milk production. Estrogen and progesterone produced during pregnancy may also impact mammary gland function. High concentrations of estrogens decrease milk yield, while progesterone does not. Although progesterone inhibits lactogenesis partly due to its ability to competitively inhibit binding of glucocorticoid to the glucocorticoid receptor, the inability of exogenous progesterone to impact lactation in many species is due to the lack of expression of the progesterone receptor during lactation. In dairy cows this inability is due to a reduction in number of progesterone receptors rather than an absolute absence of the receptors in the lactating gland. The mechanism by which estrogens decrease milk production is unclear.

Galactopoietic Effect of Milk Removal

Removal of milk from the mammary gland is necessary for maintaining lactation. This is evident from the decreased persistency of lactation with incomplete milking and conversely from the enhanced lactation persistency in response to the facilitation of milk removal by daily injections of oxytocin at milking.

Without milk removal intramammary pressure increases, blood flow to mammary tissue decreases, and a substance (or substances) that inhibits milk secretion and promotes apoptosis apparently accumulates within the alveolar milk. In the short term, such as a prolonged milking interval, accumulation of milk causes partial inhibition of milk synthesis and secretion; in the long term it causes the termination of lactation and initiation of mammary involution. Considerable research effort has been expended in attempts to identify products in milk that feed back to inhibit milk secretion. Initial research tentatively identified a substance with appropriate characteristics that was referred to as feedback inhibitor of lactation (FIL), but the substance has not been purified and identified or its gene identified. Currently, a naturally produced proteolytic fragment of casein has been identified and the synthesized peptide sequence shown to inhibit milk synthesis. These latter results are preliminary but intriguing. Additionally, milk accumulation produces leakiness of the tight junction complexes between epithelial cells. Artificially increasing tight junction leakiness also decreases milk secretion by mechanisms as yet undetermined, but they are proposed to involve signal transduction via the cytoskeleton (Stelwagen, 2001).

The benefits of frequent milking are multifaceted. Increasing milking frequency from twice daily to thrice daily increases milk production by approximately 20 percent. This benefit accrues largely because of the removal of the inhibitory effects of milk accumulation. In the short term, increased milking frequency appears to cause increased cellular activity, whereas in the long term glands milked more frequently have a greater number of cells. Because these results can be demonstrated by milking glands within the same udder at different frequencies, the effects are presumably due to local effects rather than systemic responses. Limited studies indicate that the rate of decline of milk production after peak lactation (persistency) is the same in cows milked three times and two times. Finally, increased milk production can be realized by temporarily increasing milking frequency during early lactation (Bar-Peled et al., 1995). This carryover effect suggests that the increased milking frequency induces an increase in mammary cell number and thus an effect that persists through the lactation. Conversely, it was demonstrated over 30 years ago that a temporary reduction in milking frequency during early lactation causes a decrease in milk yield that persists for that lactation. Although the effect of milking frequency appears to be largely mediated by local effects, interactions involving secretion of galactopoietic hormones that are released at each milking in addition to oxytocin are possible. These hormones include Prl, and glucocorticoids, as well as bST in goats and rats, but not in cows or humans.

Other Aspects of Galactopoiesis

As mentioned previously, increased expression of mitogenic factors and cell survival factors in the mammary gland can increase milk production by virtue of their effects on the size and maintenance of the alveolar secretory cell population. Beyond the classical hormones and growth factors already discussed, there are many agents that can impact cell turnover. A greater understanding of known factors and the potential discovery of new factors will certainly increase the opportunity to regulate persistency of lactation. Perhaps a newly discovered agent will become the next generation bST.

Chapter 8
Biochemical Properties
of Mammary Secretions

Although most discussions of milk focus on secretions obtained once lactation is established, it is useful to consider the biological significance of changes in mammary secretions. In many species colostrum formation is critical for offspring survival. A contrast between humans and cows illustrates this point. In humans the organization of the placenta (hemochorial) allows the transfer of immunoglobulins from the mother to the developing fetus in utero. This means that the human infant is born with at least a degree of passive immunity. In contrast cows and other ruminants have a synepithelialchorial type of placental barrier so that immunoglobulins do not pass to the developing offspring during fetal development. Fortunately, the initial mammary secretions or colostrum are usually rich in immunoglobulins. For a brief period postpartum immunoglobulins in the colostrum can be absorbed intact across the gut mucosa and appear in the serum. After a period of time from about 3 days to several weeks in some species, this uptake capability stops. The passive humoral immunity supplied to the neonate at this time is essential for its survival. Although colostrum does contain agents that provide some nonspecific resistance to disease (e.g., lysozyme and lactoferrin), it is the presence of specific antibodies that is critical. For example, it is known that calves reared normally on colostrum and milk from unimmunized dams quickly succumb to infections to which the dam has never been exposed. On the other hand, immunization of the dam prior to parturition also protects the calf from postnatal exposure to the same pathogen, provided the calf is given colostrum from its immunized mother. Experiments have clearly demonstrated that specific immunity provided by colostrum feeding is more important than nonspecific elements in mammary secretions. The metabolic adjustments necessary in the transition from a fetus to a newborn are dramatic. In addition to the need for immunity, neonates typically have little available fat for energy, and many of these minor stores are critical for development and therefore not readily available as an energy reserve. The limited amount of glycogen stored in the liver or other tissues is rapidly used; plus the liver capacity for gluconeogenesis is limited at this time. This means that mammary secretions are critical to supply both immunological protection and nutrients needed for survival (Davis and Drackley, 1998).

Colostrum Formation

The gross composition of bovine mammary secretions obtained during the first few days after calving is illustrated in Figure 8.1. Similar changes have been noted for secretions from goats, sheep, and humans. After the first few milkings, composition rapidly changes to that of normal milk. Apparent changes include an increase in lactose and a dramatic decrease in total protein. The loss in protein is primarily a result of the dramatic decrease in immunoglobulin content. Transport of maternal immunoglobulins into colostrum probably occurs in all mammals to a degree, but the significance of the immunoglobulins in colostrum depends on the species. In primates transport of immunoglobulins to the fetus through the placenta occurs by a receptor-mediated, intraepithelial mechanism similar to that in

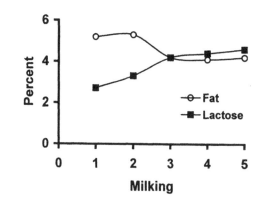

A

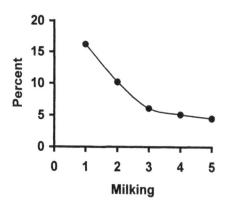

B

Fig. 8.1 Change in bovine milk composition at successive milkings following parturition is shown. **A** shows changes for lactose and fat, and **B** total protein. Data adapted from Parrish et al., 1948.

the mammary gland. This means that the human infant is born with a complement of antibodies in its blood to protect it from disease until its own immune system is activated and is fully functioning. Immunoglobulins in the colostrum of primates are nonetheless important because of likely effects in the protection of the gut mucosa as well as additional passive immunity through uptake across the gut. After ingestion of colostrum by the neonate, the antibodies are absorbed intact into the neonate's bloodstream. This unique absorption process in the intestine stops after a time postpartum that varies between species. This halt in intestinal absorption of immunoglobulins and other macromolecules is called *gut closure*.

Immunity Overview

The basis of specific immunity arose from observations in the late 1800s, when researchers demonstrated that animals that survived a bacterial infection had protective agents in their blood (now known to be immunoglobulins or antibodies) that defended the animals against attack by the same pathogen. It was also shown that if antibody–containing serum from the surviving animals was given to animals that had not been exposed to the pathogen, these animals were protected against attack as well. It was initially believed that production of antibodies was the only critical requirement for protection. However, in cases where transfer of antibody serum sometimes failed to provide protection, transfer of the donor's lymphocytes often did provide protection. As experiments evolved, it became clear that immunity involved both circulating antibodies as well as populations of leucocytes. Thus, the two overlapping arms of the immune system are humoral and cell-mediated immunity.

Briefly, two broad classes of lymphocytes are recognized. Upon stimulation B lymphocytes, usually simply called B cells, are induced to divide to create clones of cells, some of which are retained as memory B cells. Others of the cells differentiate into plasma cells that synthesize and secrete large quantities of antibodies. The presence of the memory B cells is the explanation for the marked increase in antibody concentration (titer) when animals are exposed to the same pathogen or antigen a second time. Cell-mediated immunity depends on T lymphocytes, which include several subclasses of cells (helper T cells, cytotoxic T cells, and suppressor T cells). Other cells critical to functioning of both arms of the immune system include the antigen-presenting cells (fixed and wandering macrophages) as well as neutrophils. B cells are named by the fact that they were first described as lymphocytes present in the bursa of Fabricius of birds and T cells because they mature in the thymus. In mammals B cells are believed to mature in the bone marrow. In reality the actions of the two divisions of the immune system are closely interwoven. But generally humoral immunity is most effective against bacteria and their toxins or free viruses because antibody binding can inactivate these attackers and make them susceptible to destruction by phagocytic cells or complement activation. Cell-mediated immunity is a better weapon against cellular targets;

examples include cells that have been infected by viruses, parasites, or perhaps cancer cells.

Although it has been proposed for some species that lymphocytes present in colostrum might be important in protection of the neonatal gut and even that these cells might populate neonatal tissues, there is a stronger case for involvement of humoral immunity. There are five major classes of immunoglobulins: IgA, IgG, IgM, IgD, and IgE. In ruminants nearly all of the antibodies in mammary secretions are IgGs. However, in human colostrum or milk, more than 80 percent of the antibodies are the IgA type, and in porcine mammary secretions IgA and IgG account for 15 and 75 percent of the antibodies, respectively. IgA in secretions is often linked to antibodies produced by local tissue plasma cells (rather than transport from blood plasma). This suggests the possibility that much of the protection afforded to the neonate or mammary gland (e.g., mastitis resistance) is the result of local stimulation of cells within the mammary gland in species with high concentrations of IgA in their colostrum or milk. Table 8.1 provides a comparison of the relative proportion of immunoglobulins in serum and mammary secretions of three representative species (Guidry, 1985).

While the focus of this section is on the role of mammary secretions in protection of the neonate, defense of the mammary gland against mastitis is equally important. As reviewed (Sordillo et al., 1997), much work is underway to develop techniques to control immunological defenses within the udder. In particular the availability of recombinant cytokines has stimulated a number of studies to determine if these agents are useful as therapies to treat or prevent mastitis. Other studies suggest that supplementation with specific micronutrients may also be important in enhancing local mammary immunity.

Table 8.1 Concentrations of immunoglobulins in the serum and mammary secretions of three representative species

Species	Immunoglobulin	Concentration (mg/ml)		
		Serum	Colostrum	Milk
Human	IgG	12.10	0.43	0.04
	IgA	2.50	17.40	1.00
	IgM	0.93	1.60	0.10
Bovine	IgG	18.90	50.50	0.80
	IgA	0.50	3.90	0.20
	IgM	2.60	4.20	0.05
Porcine	IgG	21.50	58.70	3.00
	IgA	1.80	10.70	7.70
	IgM	1.10	3.20	0.30

Source. Data adapted from Guidry, 1985.

Antibody Structure and Function

Whatever its specific class, each antibody molecule has a basic structure consisting of four protein chains linked together by disulfide bonds. Two identical so-called heavy chains of about 400 amino acids each make up the bulk of a structure that resembles the arms and stem of the letter Y. Two additional identical protein chains (the light chains) essentially overlap the portions of the heavy chains in the arms of the Y and are joined by sulfide bonds. The heavy chains have a hinge-like region near the top of the stem of the Y. The two ends of the Y of the antibody molecule create the sites for binding of the antibody to its antigen. This means that the antibody is divalent; that is, one molecule is capable of binding two antigen molecules. The stem region of the molecule is important because it contains sites for complement binding and for macrophage activation. These sites are important because the binding or fixing of the antibody allows the development of a cascade of reactions important in the immune attack. The structure of an antibody molecule is illustrated in Figure 8.2.

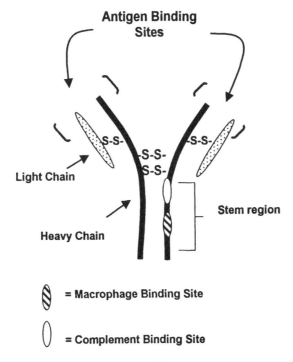

Fig. 8.2   A basic antibody molecule consists of four protein chains, two heavy and two light chains. The portion of the heavy chains resembling the stem of the letter Y provides sites for activation of complement and interactions with macrophages. The combination of heavy and light chains at the ends of the arms of the Y creates two identical antigen-binding sites. Disulfide bonds are prominent in linking heavy and light chains together as well as at locations within the chains (not illustrated).

Although antibodies do not directly destroy pathogens, their binding to antigens on the surface of bacterial cells, for example, acts to mark the cells for destruction. Antibody binding to toxins or foreign debris can inactivate these components by neutralization, precipitation, or, in the case of cell-associated antigens, agglutination. These reactions greatly enhance inflammation, and along with inflammation chemotaxis recruits leucocytes that destroy bacterial cells by phagocytosis. Antibody binding also triggers complement fixation and exposes the macrophage-binding sites on the antibody molecule. The coating of foreign substances with antibodies is called *opsonization.*

The complement system, usually simply called *complement,* is a complex of at least 20 proteins present in the circulation. These proteins provide an important mechanism for the destruction of foreign substances because, when activated, they greatly enhance the inflammatory response and even more impressive, they can stimulate the direct destruction of bacteria and some other cells by causing the cells to rupture. While the complement system is really nonspecific because it responds to many fixed or opsonized antibodies, it clearly enhances or "complements" the immune response, hence the name. Briefly, the complement system is activated by one of two pathways. In the classic reaction the binding of antibodies to the surface of a pathogenic bacterial cell allows the binding of complement protein C1 to the antigen-antibody complex. This reaction promotes a series of related protein-protein interactions that eventually cause the cleavage of complement protein C3 into two fragments (C3a and C3b). This promotes a cascade of reactions when C3b binds to the surface of the bacterial cell and a membrane attack complex is formed (a cluster of related complement proteins). This complex of proteins stabilizes an opening in the bacterial cell membrane that interferes with calcium transport in the bacterial cells. These events lead to destruction of the bacterial cell by lysis.

Antibody Transport and Receptors

Given the importance of immunoglobulins to neonatal health, a number of studies have focused on regulation of immunoglobulin transport across the mammary epithelium. In reality the IgG in mammary secretions belongs to two subclasses. IgG_1 is the major (~95 percent of the total IgG) immunoglobulin transported into colostrum in the bovine mammary gland during colostrum formation, although there is some specific transport of IgG_2 as well. Because the IgGs come almost exclusively from the blood (compared with local tissue production of IgA), understanding transport mechanisms is important. Uptake of the antibodies occurs by uptake of the molecules via endocytosis, transfer of these antibody-containing vesicles to the apical ends of the alveolar cells, and release by exocytosis. The process is receptor mediated, but there are unique receptors for the various immunoglobulin subtypes. The receptor for IgA and IgM expressed by the mammary epithelial cells is the secretory piece or component (SC). When bound by antibody, it is proteolytically cleaved from the membrane as IgA is transported across the epithelial cell. After exocytosis into the alveolar lumen, the SC remains bound to the IgA so that SC-IgA complexes (now called *secretory IgA*) appear in the secretions. Since nonbound SC is also present in milk and colostrum, this suggests that cleavage and transport of SC does not require that it be bound to IgA. It may be that expression of SC is induced at the usual time of colostrogenesis and that trafficking of the SC begins regardless of the availability of IgA for transfer.

The receptor(s) for IgG transfer is (are) less well characterized, but specific binding sites on the mammary epithelium are known. Transfer of IgG_1 into colostrum begins in late gestation but rapidly decreases with the onset of lactogenesis. The inverse relationship between lactogenesis and IgG transport suggests that hormones responsible for alveolar cell differentiation may also suppress colostrogenesis and especially immunoglobulin transfer. This idea is supported by in vitro experiments with mammary explants taken from cows at the time of colostrum formation. Specifically, mammary cell expression of the IgG_1 receptor, judged by immunocytochemistry, was reduced when explants were incubated with Prl and in conjunction with increased secretion of α-lactalbumin. This is also supported by results from hormone induction of lactation in cows. To induce lactation, cows were given subcutaneous norgestomet implants and daily injections of estradiol for 7 days. Implants were removed at this time, and daily treatments (days 8–11) with recombinant bovine Prl or bromoergocryptine (to suppress normal Prl secretion) initiated. Mammary biopsies were obtained on day 8 and 12 of the study. As shown in Figure 8.3, the concentration of IgG_1 was markedly higher in mammary secretions from bromoergocryptine-treated cows in the later phase of the induction scheme. On day 12 cows with suppressed Prl secretion expressed abundant amounts of IgG_1 receptor (68 ± 3 percent of tissue area showed IgG_1

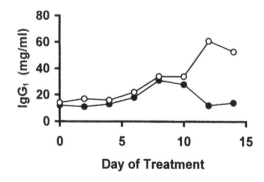

Fig. 8.3 Mean concentrations of IgG_1 in mammary secretions of cows induced into lactation. Note the divergence in immunoglobulin concentrations in cows with normal Prl (*closed symbols*) compared with those treated with bromoergocryptine (*open symbols*) to suppress secretion of Prl. This suggests that lactogenesis as associated with enhanced Prl secretion suppresses colostrum formation and specifically immunoglobulin transport into mammary secretions. Data adapted from Barrington et al., 1999.

receptor staining), but there was almost no expression of the receptor in tissue from controls or cows given Prl. At this time Prl concentrations were negatively correlated with mammary secretion IgG_1 concentration ($r = -0.79$) and with IgG_1 receptor score ($r = -0.77$). Concentrations of IgG_1 in mammary secretions were highly correlated with the tissue IgG_1 receptor score ($r = 0.96$). These data strongly support the idea that, in addition to its importance in promoting lactogenesis, Prl secretion near the time of calving also signals the cessation of colostrogenesis (Barrington et al., 1999).

Colostrum Feeding and Management

It is difficult to overstate the importance of colostrum feeding in ruminants. The first meal of colostrum provides not just critical immunoglobulins but other essential nutrients as well. Compared with normal milk, the concentrations of total protein and fat are much greater. In addition a variety of other nutrients are more concentrated in colostrum (e.g., zinc, iron, folic acid, choline, riboflavin, and vitamins, A, E, and B_{12}). The relative importance of these nutrients and the need for energy production are clear if you compare a calf and a human neonate. The body of the newborn calf contains about 3 percent lipid, compared with about 16 percent for the human infant. Moreover, much of the lipid in the calf does not come from adipose depots that can be stimulated to supply fatty acids, since many of the lipids are structurally related. It is estimated that the total reserves of fat and glycogen available to the calf at birth would be used completely within about 18 hours without supplemental feeding (Davis and Drackley, 1998).

With the advent of modern dairy practices, producers began to remove calves from their dams shortly after birth to improve the efficiency of management practices on their farms. This was accompanied by dramatic increases in enteric infections and general difficulties successfully raising calves. Despite the presence of supplements in calf diets (often antibiotics), problems remained. These observations made it clear that calves (and other ruminants) depend on colostrum to provide the antibodies necessary for systemic and gut-associated immunity in the first few weeks of life. Colostrum can be a very rich source of immunoglobulins. Dairy cows, for example, can secrete as much as 2 kg of IgG in the colostrum obtained during the first 4–5 milkings postpartum. On the other hand, colostrum from individual cows can vary dramatically. Figure 8.4 illustrates the distribution of IgG_1 concentrations in colostrum from a large dairy farm. It is now a routine practice to preserve or freeze any extra colostrum so it can be fed to calves with problems or to supplement colostrum production for some dams.

A generally accepted measure of the success of colostrum feeding is for the calf to attain a plasma or serum IgG concentration ≥ 10 mg/ml by 48 hours postpartum. A goal for better management is probably closer to 15 mg/ml. To meet this goal, it is essential to begin colostrum feeding as soon as possible after parturition since the capacity for transfer across the gut disappears quickly and may be hastened by hypothermia, dystocia, or other difficulties. Also, a good quality colostrum must be

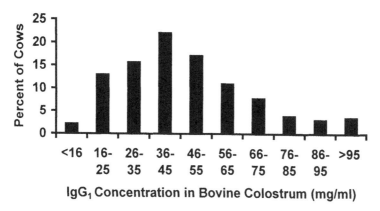

Fig. 8.4 Distribution of IgG$_1$ concentration in colostrum from Holstein cows is illustrated. Data adapted from Gay, 1994, and Davis and Drackley, 1998.

fed. A good quality colostrum contains ~60 mg/ml of IgG. If the concentration of IgG is lower, then a larger volume or more frequent feeding would be needed to achieve the same result. A standard recommendation is to feed 1.89 L (2 qt) of high-quality colostrum within the first few hours of life, followed by a second feeding within 12 hours. However, if the goal is to supply 100 g of IgG to the calf, calculations based on the distribution illustrated in Figure 8.4 show that only 29 percent of the calves would receive this quantity of IgG at the first feeding. Increasing the feeding volume to 2.82 L (3 qt) increases the percentage to 71 percent, and increasing the feeding volume to 3.78 L (4 qt) to 81 percent of the calves. Although neither lower nor higher IgG concentrations guarantee that there will be no illness or health problems, mortality rates are lower in operations that carefully monitor colostrum feeding. Figure 8.5 illustrates survival curves for calves with adequate (>10 mg/ml) compared with calves with inadequate (<10 mg/ml) serum concentrations of IgG in their serum. Fortunately, since a very large portion of the serum protein of the neonatal calf is accounted for by the immunoglobulin content, measurement of total protein (via a refractometer or other simple chemical means) provides a good measure of immunoglobulin absorption (Quigley and Drewy, 1998).

Milk Yield and Composition During Lactation

Compared with the dramatic changes in the composition of mammary secretions around the time of parturition, there is relatively little change in milk composition during most of the lactation. However, a variety of factors (e.g., feeding or nutrition, season, breed, milking frequency, environmental stresses, or metabolic disturbances) do impact milk yields and milk composition to a lesser degree. The concentration of milk fat is the most changeable of the major milk components during established lactation. The following discussion focuses on bovine, but by extension analogous effects occur in many lactating mammals.

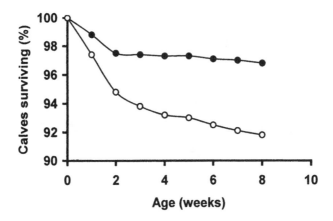

Fig. 8.5 Survival of calves with adequate (>10 mg/ml, *closed circles*) and inadequate (<10mg/ml, *open circles*) immunoglobulin concentrations in serum is shown. Data from the National Dairy Heifer Monitoring Evaluation Project completed by the National Animal Health Monitoring Systems, NAJMS, 1993; adapted from Davis and Drackley, 1998.

Many potential problems are relatively easy to predict. For example, onset of milk secretion is tremendously energy and nutrient demanding. Consequently, if the lactating cow is not able to increase mobilization of tissue stores and/or adjust feed intake to account for the demands of lactation, metabolic problems such as ketosis or hypocalcemia can occur. Problems can be exacerbated if dairy managers fail to anticipate nutrient needs and don't feed their animals adequate amounts of balanced rations. Other alterations in milk composition can be relatively subtle and relate to the consumer's perception of flavor, texture, or possibly odor. Fat is the most variable milk constituent, and minerals (ash) and lactose the least variable. Some breeds of cows are also noted for a difference in milk composition, but differences among individuals within a breed may be even greater than differences among breeds. Some general nutritionally mediated effects on milk composition occur with changes in the proportions of concentrates and forages in the diet. For example, high-grain feeding, especially if accompanied with reduced forage intake, can act to change rumen fermentation patterns so that the relative amount of propionate and acetate produced is changed. An increase in the propionate: acetate ratio favors reduced milk fat synthesis and therefore a reduction in milk fat percentage. Milk fat depression can also occur by feeding lush forages or feeding normal forages that are too finely chopped. A typical lactating cow diet contains 15–17 percent crude fiber to support a rumen fermentation pattern conducive to milk fat synthesis. Interestingly, the diameter of the milk fat droplet varies considerably from about 1 to 10 µm within animals as well as between breeds. For example, Guernsey cows have the largest fat droplets, and Holsteins and Ayrshires the smallest milk fat globules. A general rule is that the higher the fat percentage, the larger the

Table 8.2 Composition of milk from five dairy cattle breeds

	Percentage by weight					
Breed	Total solids	Fat	Casein	Whey protein	Lactose	Ash
Ayrshire	12.7	3.97	2.68	0.60	4.63	0.72
Brown Swiss	12.7	3.80	2.63	0.55	4.80	0.72
Guernsey	13.7	4.58	2.88	0.61	4.78	0.75
Holstein	11.9	3.56	2.49	0.53	4.61	0.73
Jersey	14.2	4.97	3.02	0.63	4.70	0.77

Source. Adapted from Jenness, 1985.

diameter of the fat particle. Also, its size usually decreases as the stage of lactation advances. Another breed difference example is carotene. Carotene is a precursor of vitamin A and appears as a yellow pigment. Because Guernsey and Jersey cows convert less carotene to vitamin A than other breeds of dairy cattle, milk from Guernsey and Jersey cows has a distinctive yellow color. This explains the expression "golden Guernsey" or "Jersey milk." Differences in milk composition between dairy breeds are shown in Table 8.2.

At parturition milk production begins at a high rate, but production continues to increase for about 3 to 6 weeks. Higher-producing cows usually take longer than low-producing cows to attain their peak milk production. After this time milk production gradually declines. The rate of decline is referred to as *persistency.* Cows with a slower rate of decline are described as more persistent. As discussed in Chapter 7, bST treatment, for example, increases the persistency of lactation. Typical rates of decline would be about 6 percent per month for cows in the first lactation and about 9 percent per month for multiparous cows. Nonpregnant cows can continue to secrete milk indefinitely, but yields after prolonged periods are typically very low.

Milk composition also changes during lactation. For example, the milk fat percentage typically decreases slightly during early lactation but then increases toward the later period of lactation. Milk protein concentration gradually increases with advancing lactation. Lactose and mineral concentrations may also increase slightly during advanced lactation. Much of the increase in these components of milk is associated with advanced stages of concurrent pregnancy rather than stage of lactation.

Off-flavors can appear in milk from feeding of some pasture plants (e.g., garlic, onions, and even some silage mixtures). These appear because of the absorption of volatile oils present in the plant material and the transfer of these substances from the blood into milk. It is also possible for contaminants to get into milk during storage, transport, or processing. The following sections briefly describe some of the variables known to impact milk production or milk composition.

Bioactive Milk Peptides, Growth Factors, and Hormones in Milk

Although the normal composition of milk with respect to its general nutritional value is well appreciated, other biological properties of milk are not. Like many other biological secretions, normal milk contains a large number of naturally occurring biologically active hormones, growth factors, and other peptides. As recently reviewed (Clare and Swaisgood, 2000), the proteins in milk can provide the starting material to generate many other so-called bioactive peptides. Some of these occur naturally as a part of the usual digestive process in the gut of the suckling offspring, but others can be purposefully manufactured. Some these manufactured peptides could greatly improve the safety of milk-derived foods by preventing or minimizing the effects of contaminating microorganisms. Others could fill important niches as value-added components to improve the health-related properties of dairy foods.

Hormones and Growth Factors in Milk

As a general rule since milk is produced and stored in the alveolar and ductular lumen within the mammary gland and interstitial fluids derived from the blood supply precursors to the mammary cells, it should come as no surprise that many substances found in blood or interstitial fluids can also be found in milk. This does not mean that there is necessarily equilibrium between components in blood and milk, but rather it emphasizes that milk is a biologically derived nutrient source, and thus the presence of these hormones and growth factors is completely natural. As endocrine-related assay methods became more sensitive from the late 1970s to the current time, the list of hormones and growth factors known to be present in milk has grown progressively (Pope and Swinburne, 1980; Grosvenor et al., 1993). In many cases the appearance of a particular hormone in milk correlates with the expression of receptors for the hormone by the mammary secretory cells.

In the bovine, the pituitary hormone Prl is routinely detected in both milk and colostrum. Prl concentrations in secretions obtained in the periparturient period are typically greater in mammary secretions than in serum. Near the time of calving, Prl concentrations in secretions average about 300 ng/ml, but in established lactation concentrations are typically 30–50 ng/ml. In fact, the ratio between secretion/plasma averages about 8 in the week before calving, remains at about 4 between –1 and 3 days postpartum, then approaches approximately 1 after the first week of lactation, and remains at this level thereafter (Malven, 1977). Moreover, it is generally accepted that much of the Prl in milk is derived from receptor-mediated uptake from maternal circulation. In established lactation in the cow, changes in average concentrations of serum Prl are positively correlated with milk Prl concentrations. It is also known that transfer of Prl into milk appears to occur more readily at the time of milking and that microtubules are involved in the transport (Akers and Kaplan, 1989). However, in the humans and rats, for example, Prl concentrations determined by radioimmunoassay are generally much less than for concentrations measured by the Nb2 bioassay technique (see Chap. 6). Furthermore, three classes of Prl variants are found in the milk of these species: phosphorylated, glycosylated, and soluble receptor or binding protein bound. This is relevant because in humans about 70 percent of the Prl in milk is the phosphorylated form. Alkaline phosphatase treatment of milk-derived human Prl more than doubles the Prl activity detected by the Nb2 bioassay but has no effect on the quantity detected by radioimmunoassay. To further complicate the situation in humans and rats, it is also known that the low levels of Prl mRNA are expressed in the mammary gland. The impact of possible local expression of Prl on concentrations of Prl in mammary secretions is unknown (Ellis et al., 1996).

In contrast with Prl, very little GH is detected in milk, despite the fact that these hormones are very similar in structure and basic chemical properties. This concentration seems to be related to the apparent low level of specific receptors in the mammary epithelium for GH. This is further indicated by the lack of a detectable difference in the concentration of GH in control cows compared with those treated with bST during lactation. This is despite the fact that bST-treated cows have consistently elevated concentrations of GH in their blood following treatment with prolonged release formulations of the hormone. The same is generally true for concentrations of IGF-I in milk. Specifically, IGF-I accumulates in mammary secretions (as much as 1000 ng/ml) around the time of calving, but once lactation is established, concentrations in milk rapidly decrease and are maintained at a low concentration of 5–10 ng/ml throughout lactation. Cows treated with bST typically have elevated concentrations of serum GH from baseline concentrations of 2–10 ng/ml to concentrations of 35–75 ng/ml (depending on the sampling day relative to treatment with the prolonged release product and stage of lactation) and typically about a doubling in plasma IGF-I. Yet changes in absolute concentrations of milk GH and IGF-I are minor in comparison. For example, Zhao et al. (1994) detected short-lived spikes in milk GH for cows given slow release injections (~1.4 ng/ml) and average concentrations of 0.6, 0.8, and 1 ng/ml for cows given a placebo, daily bST injections, and the slow release formulation every 2 weeks. Mean concentrations of milk IGF-I for the same treatments were 4.31, 4.82, and 4.52 ng/ml. In this trial, the difference between control cows and cows given bST was statistically significant only for cows given daily injections of bST. IGF-I in milk also varies depending on stage of lactation (i.e., IGF-I in milk averaged 3.5, 4.3, and 6.1 ng/ml for control cows in early, mid-, and late lactation, respectively). Collier et al. (1991) previously reported that milk IGF-I was affected by stage of lactation, parity (higher in multiparous cows), and farm of origin. Concentrations of IGF-I in milk were not normally distributed either. The range of values above the median exceeded those below the median. A more recent study (Daxenberger et al., 1998) reported similar results (i.e., effects of parity and stage of lactation), but the researchers found small but statistically significant increases in milk IGF-I for cows treated with bST. They suggested that measurement of milk IGF-I might be useful as a tool to monitor cows treated with bST, but this seems problematic since values for cows treated with bST often fall within the range of concentrations found in the milk of untreated cows. Indeed, the overall conclusion from their study, despite the introduction of some improvements in the assay performance, supports the ideas outlined above: "measurements in milk samples might be useful to monitor bST treatment, although they can not be used for forensic purposes. The technique presented in this paper can be applied as a screening procedure prior to further confirmatory testing, for which commonly accepted methods do not yet exist." In summary it seems prudent to remember that even though the concentrations of IGF-I in milk of cows given bST are increased, these values are well within the range for untreated cows.

It has also been noted that preparation of bovine milk for use in infant formula (heated to 121°C for 5 minutes) makes IGF-I undetectable by radioimmunoassay. Groenewegen et al. (1990) found that feeding of milk from control or bST-treated cows or additions of purified bST to milk of control cows (500 ng/ml) had no ability to stimulate the growth of hypophysectomized male rats. However injections of hypophysectomized rats with 5, 20, or 80 µg/day of bovine GH produced the expected dose-related increase in average daily gain and width of the tibial epiphyseal growth plates. These data suggest that treatment of cows with bST has only a minor effect on concentrations of GH or IGF-I in milk and, as judged by a standard growth assay protocol, no effect on the biological activity of the milk. The concentration of IGF-I in bovine milk is typically lower than in human milk; other possibly potent growth factors (e.g., EGF) are markedly higher in human milk. This does not mean that there should be no consideration of possible effects of protein hormones and peptide growth factors in milk or dairy products destined for human consumption—only that changes in milk composition should be considered within a relevant context and physiological framework.

Steroid hormones, on the other hand, can readily pass across the epithelium and appear in milk. Depending on the biochemical properties of the specific steroids, these hormones tend to associate with the milk lipids rather than the aqueous phase of the milk unless associated with specific binding proteins. Indeed, measurement of milk progesterone can be used as a screening tool to determine the pregnancy status of cows during lactation. Other nonpeptide hormones found in milk include thyroxine, triiodothyronine, reverse triiodothyronine, cortisol (or other glucocorticoids), and various estrogens. It is generally accepted that the nonpeptide hormones can be absorbed from the gastrointestinal tract of humans and other mammals. This suggests that changes in this class of milk-derived hormones *could* impact both neonates and older animals or humans consuming milk product–containing diets.

Irrespective of the great amount of research and commentary associated with the introduction of bST as a tool to enhance milk production in dairy cows, the physiological relevance of some growth factors found in milk to the developing neonate is better understood. A number of growth factors are increased in colostrum (in a pattern that mimics the accumulation and subsequent decline of immunoglobulins), which has led a number of researchers to hypothesize that the accumulation of growth factors in colostrum must have evolved to promote neonatal growth or development (Koldovskỳ, 1996). One of the most extensively studied of these growth factors is EGF. It is naturally present in the milk of rodents, rabbits, and humans but not in the milk of cows. When EGF was added to milk substitutes given to suckling rats, those receiving the EGF had an enhancement of intestinal tract development. More telling, when pups were fed pooled rat milk containing antibodies against EGF, intestinal tract development was impaired compared with that of controls. These observations support the idea that at least some of the growth factors in milk may have important physiological roles in the physiology of the neonate. Table 8.3 provides a list of peptide hormones and growth factors present in human milk.

Table 8.3   Human milk–derived peptide hormones and growth factors.

Hypothalamic-hypophyseal hormones
Gonadotropin-releasing hormone (GnRH)
Growth hormone–releasing hormone (GRF)
Growth hormone (GH)
Prolactin (Prl)
Thyrotropin-releasing hormone (TRH)
Thyroid-stimulating hormone (TSH)

Thyroid-parathyroid–related peptides
Calcitonin-like peptide
Parathyroid hormone-like peptide

Gastrointestinal-related peptides
Gastrin
Gastric inhibitory peptide (GIP)
Gastric-releasing peptide (GRP)
Neurotensin
Peptide histidine methionine (PHM)
Peptide YY
Somatostatin
Vasoactive intestinal peptide

Growth factors
Epidermal growth factor (EGF)
Transforming growth factor α (TGF-α)
Transforming growth factor β (TGF-β)
Insulin-like growth factors I and II (IGF-I; IGF-II)
Insulin
Neural Growth Factor (NGF)

Source. Data adapted from Koldovský, 1996.

Specifically related to calves, as reviewed by Baumrucker and Blum (1994), calves given colostrum have higher blood concentrations of IGF-I than those given milk or milk replacer. It was also shown that radiolabeled IGF-I added to the diet was absorbed and the intestinal mucosa of the newborn expressed IGF-I receptors. A subsequent study was designed to compare the effects on calves fed diets with milk replacer alone, pooled colostrum for 4 days followed by milk replacer, or milk replacer supplemented with 750 ng/ml IGF-I for 1 week on serum IGF-I concentrations. Although pilot studies with newborn calves showed the addition of 1.125 or 6.25 mg of recombinant IGF-I fed in 200 ml of buffer doubled serum IGF-I concentrations within 4 hours, dietary treatments had no effect on serum IGF-I during the first 4 days of treatment. It was hypothesized that the lack of appearance of IGF-I in the serum of calves given the supplemented diet might reflect the effects of dietary proteins or fats compared with IGF-I added to buffer alone. Regardless, addition of exogenous IGF-I to colostrum or milk replacer would likely better mimic naturally occurring exposure of the suckling neonate to IGF-I in mammary secretions. These data give little support to the notion that addition of exogenous growth factors to milk replacer or colostrum can be easily used to modulate the health or productivity of the calf—especially since the absorption of intact peptides and proteins across the gut is expected to be greater in the neonate than in older animals. However, although total IGF-I concentration in the blood did not show any diet-induced changes in the first 4 days, a statistically significant increase was noted after this period in the calves given milk replacer supplemented with IGF-I. In addition, there were diet related differences in the secretion of insulin, Prl, and triiodothyronine, especially with the initial feedings. Thus, there is much to be learned about the possible effects of suckling and diet interactions in the endocrine and nutritional physiology of the neonate (Blum and Hammon, 1999).

It is likely true, but not proven, that most of these same agents also appear in the milk of cows and other dairy animals. However, it is equally likely that these hormones and growth factors would have little impact on the majority of the milk- or dairy product–consuming public. Two general observations support this conclusion. Except for the possibility of de novo generation of bioactive peptides from milk proteins, clearly the larger protein hormones are likely digested along with most other food-derived proteins. Second, concentrations of the majority of hormones and growth factors in mature milk are typically much lower than in blood. Unless consumers are on an all-milk or -dairy diet, simple bulk dilution will markedly reduce effective concentrations of any of these agents in the lumen of the GI tract, further reducing any opportunity for uptake. The processing of milk is also more likely to reduce biological activity than to increase activity of these naturally occurring peptides. Since the suckling young have the capacity to absorb the large molecular weight immunoglobulins, it is evident that the absorptive properties of the neonate digestive tract is much more likely to allow absorption of proteins and peptides than the more mature animal digestive tract. Indeed, the time of gut closure (the time postpartum when the capacity for immunoglobulins to appear in neonatal circulation stops) varies considerably across species. This suggests that absorption of essentially any of the milk components is probably greatest in the early postnatal period.

However, it is also likely true that at least some of these agents, especially those in colostrum, evolved to impact intestinal tract function or perhaps other physiological functions in the neonate. This means that human milk, like that of other mammals, contains a number of hormones, growth factors, and other bioactive agents. As discussed in a review by Koldovský, (1996), some have suggested that hormones should actually be added to formula prepared for human infants to better mimic "natural" human milk. Common sense would then suggest that in most cases the nutrition of newborns is best served by feeding of mother's milk rather than formulas derived from the milk of other species—or other nutrients for that matter.

Bioactive Peptides

Although the appearance of known hormones or growth factors in milk could be logically expected to impact the biological properties of milk, a variety of bioactive peptides can also be generated from hydrolysis of milk proteins. Some of these agents are naturally produced as a consequence of usually digestive processes, and others are generated by in vitro enzymatic hydrolysis of purified milk proteins for manufacturing purposes (Schanbacher, et al., 1997; Meisel, 1997; Clare and Swaisgood, 2000). Table 8.4 provides a list of bioactive peptides that can be derived from milk.

Essentially all of the major milk proteins, as well as some nonmammary-derived proteins found in milk (serum albumin), can serve as a source of biologically activity peptides. Some of these are generated via gastrointestinal processes; others have been produced in the laboratory. Once produced in the lumen of the intestinal tract, most of these peptides must pass across the mucosal lining to produce a physiological response. Following initial identification of suspected bioactivities, many of these peptides have been purified from protein hydrolysates and assayed for bioactivity. Many of these have subsequently been chemically synthesized to confirm biological activity linked with specific peptide sequences. The biological relevance of these peptides as true physiological regulators is not established, but it is clear that possibilities are numerous. To have naturally occurring effects in vivo, the peptides must be freed during digestion and then affect target cells either within the gut lumen or after absorption across the mucosa.

Antimicrobial activity of intact lactoferrin has been known for many years, but the total antibacterial activity of milk is often not explained by the concentrations of lactoferrin alone. Recent research activities have renewed interest in the possibility that bactericidal peptides prepared from milk proteins can be useful in human or animal health. For example, the peptide casecidin is prepared by digestion of casein with chymosin or chymotrypsin and shows in vitro activity against

Table 8.4 Examples of physiologically active peptides derived from milk proteins.

Name	Amino acid	Activity
α_{s1}-Casokinin-5	α_{s1}-CN (f 23-27)	ACE inhibitor
β-Casokinin-7	β-CN (f 177-183)	ACE inhibitor
α-Lactorphin	α-LA (f 50-53)	Opioid agonist
β-Lactorphin	β-LG (f 142-148)	ACE inhibitor
Antihypertensive peptide	β-CN (f 169-174)	Antihypertensive
Casoplatelin	κ-CN (f 106-116)	Antithrombotic
Thrombin inhibitory peptide	κ-CN glycomacropeptide (f 112-116)	Antithrombotic
Thrombin inhibitory peptide	Lactotransferrin (f 39-42)	Antithrombotic
Caseinophosphopeptide	α_{s1}-CN (f 59-79)	Ca binding
Immunopeptide	β-CN (F 191-193)	Immunostimulation
Lactoferricin B	Lactoferrin (f 17-41)	Antimicrobial
β-Casokinin-10	β-CN (f 193-202)	Immunostimulation
α-Casein exorphin	α_{s1}-CN (f 90-96)	Opioid agonist
Serorphin	BSA (f 399-404)	Opioid agonist
β-Lactorphin	β-LG (f 102-105)	Opioid agonist
Casoxin C	κ-CN (f 25-34)	Opioid antagonist
Casoxin D	α_{s1}-CN (f 158-164)	Opioid antagonist
Lactoferroxin A	Lactoferrin (f 318-323)	Opioid antagonist

Source. Data adapted from Clare and Swaisgood, 2000.
Note. CN = casein; LG = lactoglobulin; BSA = bovine serum albumin; ACE = angiotensin-converting enzyme; LA = lactalbumin.

species of *Staphylococcus, Sarcina, Streptococcus,* and others. Casocidin-I, also prepared from casein, inhibits the growth of *Escherichia coli* and *Staphylococcus carnosus.* Another fragment from the N-terminus of α_{s1} casein was shown to protect mice against *Staphylococcus aureus* and *Candidia albicans.* Antimicrobial peptides from lactoferrin are active against both Gram-positive (*Bacillus, Listeria,* and *Streptococcus*) and Gram-negative (*Escherichia coli, Klebsiella, Salmonella, Proteus,* and *Pseudomonas*) species in vitro. There is evidence for the presence of lactoferricin in the human stomach after consumption of lactoferrin but the in vivo effectiveness of the peptide is unclear.

The antihypertensive peptides associated with milk act by inhibiting the activity of angiotensin-converting enzyme (ACE). ACE normally acts to cleave angiotensin I, allowing its conversion to angiotensin II, which acts to increase blood pressure directly by vasoconstriction and indirectly by promoting the secretion of the adrenal hormone aldosterone. The ACE inhibitors in milk are also produced from several of the caseins (Table 8.4). A seven–amino acid peptide, corresponding to a segment noted in hydrolysates of β-casein, has been reported to produce antihypertensive effects in a line of genetically hypertensive rats following oral administration of the peptide.

The casoplatelins are produced from the C-terminal end of κ-casein and act by inhibiting ADP-stimulated aggregation of platelets and associated binding of fibrinogen. The mechanism for these effects is likely related to molecular similarities between blood colt formation and production of casein clots in the GI tract or in cheese making. Two antithrombotic peptides noted in both human and bovine milk have been detected in the serum of newborn human infants after breastfeeding or nursing of formulae based on cow's milk.

The casein phosphopeptides are derived from α- or β-caseins after trypsin treatment. These peptides are believed to be important in forming clusters, which act as carriers for minerals but especially calcium. The peptides are resistant to enzymatic hydrolysis in the intestinal tract and are usually in a complex with calcium phosphate. These complexes improve absorption of calcium by increasing solubility. For example, when compared with animals fed soy-based diets, those fed casein-based diets have improved absorption. The casein phosphopeptides can also inhibit production of dental caries by impacting calcification of the dental enamel. Thus, their use as dental disease or prophylactic treatments has been proposed.

Milk-derived peptides affecting immune function are usually defined by responses of various blood cells. For example, immunopeptide (Table 8.4) and other small fragments from α_{s1}-casein increase the phagocytosis of sheep red blood cells by peritoneal macrophages isolated from mice. Another peptide from β-casein (amino acids 193–209), which also contains β-casokinin, stimulates the proliferation of lymphocytes in rats. Others (β-casokinin-10 or β-casokinin-7) can either inhibit or stimulate proliferation of lymphocytes depending on concentration. Mechanisms of action for these peptides on immune cells are not well understood but inhibition of proliferation of human colonic lymphocytes in response to β-casomorphin-7 apparently involves the opioid μ receptors.

Opioid peptides, or endorphins, in the body are produced from proenkephalin, propiomelanocrotin, or prodynorphin, but they share a common N-terminal sequence (Y-G-G-F). Opioid peptides from milk proteins (α- and β-caseins, α-lactalbumin, or β-lactoglobulin) also have morphine-like activity. But most of these have an N-terminal tyrosine residue and a second aromatic amino acid (phenylalanine or tyrosine) in the third or fourth position. This structure allows the molecules to complex with the opioid receptor. Differences in relative potency of the different milk protein–derived opioid peptides are associated with structural variations and consequent effects on receptor binding. Although it is natural to assume that effects produced by these peptides involve neural effects, many of the peptides are more likely to affect physiological function by directly affecting gastrointestinal tract physiology. Casomorphins are believed to prolong gastrointestinal transit time by inhibiting gut motility, producing antidiarrheal effects, altering amino acid transport, and inducing secretion of insulin and somatostatin. In contrast a precursor of β-casomorphin has been reported in the plasma of newborn calves and infants after feeding of bovine milk. It is believed that the peptide may produce an analgesic effect to cause drowsiness and sleep in infants. Other effects can be produced by direct injections of purified preparations of the various casomorphins into test animals, but whether sufficient quantities of the peptides can normally enter systemic circulation following dietary ingestion is unclear.

Related peptides, the casoxins, can act as opioid antagonists. For example, purified peptides generated by laboratory digestion of κ-casein have been shown to bind to opioid receptors in the rat brain. Specifically, if the peptides are methylated at the C-terminal end of the molecule, it selectively inhibits μ- and κ-opioid receptors in the brain. Casoxins A and B have been synthesized, and activity responses shown to correspond to amino acid sequences of peptides produced by enzymatic digestion of both human and bovine κ-caseins. However, chemically modified peptides can be markedly more active than the naturally occurring peptides produced from enzymatic hydrolysis as noted for methylated derivatives of the casoxins.

These developments might make it possible to identify dairy products containing natural health-enhancing peptides or to purposefully enhance certain products as nutraceuticals to benefit particular segments of consumers. Nutraceutical research, because of the nonnutritive nature of bioactive peptides or other substances in foods, is a growing area of interest to academic and industrial scientists. Clearly, several of the bioactive peptides identified in dairy products could be produced on a large scale for clinical use to treat diarrhea (casomorphins), mineral deficiencies (casein phosphopeptides), hypertension (casokinins), or immunodeficiency (immunopeptides). In addition to direct supplementation, it is also possible that desirable bioactive peptides could be produced during product processing by addition of enzymes or perhaps genetically engineering microorganisms. Regardless, it may be that consumers have been essentially medicating themselves for years simply by the choices made in the foods they consume. Continued research will make it possible for consumers to make informed choices about food consumption, especially in affluent Western societies, and enable health professionals to better treat their patients. Given the industrial-scale processing that has been developed to recover intact, nondenatured whey proteins in cheese making, it is also feasible to purify protein fractions to supply many of these bioactive peptides. It is reasonable to speculate that at least some of these peptides might be useful as dietary supplements, as natural preservatives for foods, or in production of specialized nutraceutical products. Many discoveries remain to be made, but the view that mammary secretions are only important as a nutrition source is rapidly changing.

Chapter 9
Management and Nutritional Impacts on Mammary Development and Lactational Performance

Interactions Between Metabolism and Intake

Given the importance of adequate nutrition to support the demands of lactation, it should be no surprise that dairy nutritionists are critically concerned with factors that control voluntary intake of dairy cows. Most of these nutritional studies focus on regulation of dry matter intake (DMI) as a tool to compare the large varieties of feedstuffs that can be used by ruminants. Critical elements can be broadly organized by considering dietary components as sources of energy (fat or carbohydrate sources sometimes referred to as *concentrate*), protein (including non-protein nitrogen, to take advantage rumen microorganisms), fiber (for maintenance of rumen health and function), and trace elements or minerals. If a properly balanced diet is prepared, measurement of the voluntary dry matter intake (VDMI) of the cow can be used to predict possible milk yield. In a pragmatic sense this means that techniques to increase VDMI are likely to increase the milk production and consequently profitability. However, as producers and nutritionists alike appreciate, VDMI is regulated in a complex fashion by factors specific to diets, management conditions, housing, behavior, environment, metabolism, and body condition.

Paradoxically, just at the time when the cow needs the most nutrients, there is routinely a decline in VDMI in the periparturient period. This is illustrated in Figure 9.1. This decline begins in late lactation and continues into early lactation. For modern dairy operations, management of these so-called transition cows is critically important. For example, most of the health problems, both of a metabolic or infectious nature, occur in early lactation. The typical decline in VDMI coincides with marked changes in reproductive status, body fat status, and the dramatic metabolic adjustments necessary to support energy, protein, and mineral demands of milk secretion. Just on the basis of energy needs, the changes are staggering. For example, it is estimated that fetal development demands on day

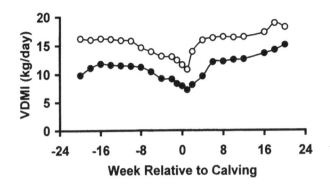

Fig. 9.1   Pattern of voluntary dry matter intake (VDMI) in heifers and cows before and after calving is shown. Data adapted from Ingvartsen and Andersen, 2000. Note the higher overall intake in multiparous cows (*open circles*) compared with that of primiparous heifers (*closed circles*) but the similarity in pattern relative to parturition.

250 of gestation (~3 weeks before calving) average 2.3 Mcal/day. The requirement for the lactating cow producing 30 kg of milk per day is estimated at 26 Mcal/day. Eating behavior and intake result from multiple interactions between neural inputs associated with the feed, feed presentation, management, metabolic conditions, and endocrine signals that are poorly understood but especially so in ruminants. Dramatic changes in VDMI occur both within and between lactations in dairy cows. Pregnant dairy heifers, for example, begin to progressively reduce their VDMI beginning several weeks before calving approximately 0.17 kg per week until 3 weeks before calving. For primiparous and multiparous cows given diets of constant composition, milk yield typically peaks at about 6 weeks postpartum, but maximum intake is not achieved until 8 to 22 weeks postpartum. Indeed, the demands of lactation require that the high-producing cow mobilize body tissues through much of the first one-third to one-half of lactation so that the animals are in a prolonged period of negative energy balance. Difference in the rate of intake recovery postpartum depends on the diet fed in early lactation as well as the degree of fatness or body condition score (BCS) at the time of calving. The normal feeding behavior is also markedly impacted by both clinical and subclinical infections so that appetite and performance is reduced in sick animals.

Hormone and Metabolic Adjustments to Support Lactation

As championed in a review by Bauman and Currie (1980) and emphasized recently by Ingvartsen and Andersen (2000), onset of lactation in high-producing dairy cows requires a coordinated, physiologically mediated reallocation of biochemical resources—homeorrhesis—to allow high milk production while maintaining homeostasis. Because of the premium placed on glucose to supply precursors for lactose synthesis and the general energy requirements of the udder, changes in circulating nonesterified fatty acids (NEFA) and glucose are especially dramatic at calving. As illustrated in Figure 9.2, concentrations of NEFA immediately postpartum are dramatically increased while glucose in the blood is reduced. This reflects the mammary demand for glucose and the corresponding stimulation of lipolysis and use of lipids as an energy source. Table 9.1 provides a partial listing of metabolic adjustments that accompany the onset of lactation.

Certainly, some of these adjustments are mediated by changes in secretion of hormones. As discussed in relation to galactopoietic effects of bST, a natural adjustment in the dairy cow, irrespective of the use of bST as a management tool, is increased secretion of GH in early lactation concomitant with a decrease in secretion of insulin. As lactation progresses, the situation is reversed (Fig. 9.3). These changes reflect the negative energy balance of early lactation and corresponding progression to positive energy balance in later lactation. The regulatory changes are mediated by more than simple changes in the secretion of GH and insulin; there are also marked adjustments in the sensitivity of metabolically critical target tissues. Table 9.2 provides a summary of changes in secretion of selected homeorrhetic and homeostatic hormones and in tissue sensitivity at mid- and late gestation compared with early lactation.

Table 9.1 Summary of major metabolic adjustments associated with the onset of lactation in high-producing dairy cow

Physiological process	Biochemical adjustment	Tissue affected
Milk synthesis	↑ Synthesis	Mammary gland
	↑ Blood flow	
	↑ Nutrient uptake	
Lipid metabolism	↑ Lipolysis	Adipose tissue
	↓ Lipogenesis	
Glucose metabolism	↑ Gluconeogenesis	Other body tissues
	↓ Glucose utilization	
Protein metabolism	↑ Proteolysis	Muscle and other tissues
Mineral metabolism	↑ Absorption	GI tract
	↑ Mobilization	Bone
Intake	↑ Food consumption	Nervous system
Digestion	↑ GI tract hypertrophy	GI tract and liver
	↑ Capacity of absorption	

Source. Data adapted from Bauman and Currie, 1980, and Ingvartsen and Andersen, 2000.

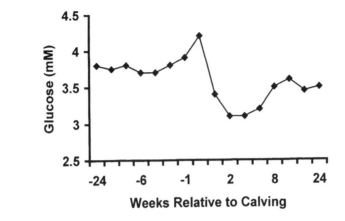

A

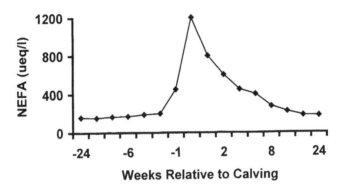

B

Fig. 9.2 Changes in plasma glucose (**A**) and nonesterified fatty acids (NEFA) (**B**) in heifers in the period before and after calving. Data adapted from Ingvartsen and Andersen, 2000.

Factors Controlling Voluntary Dry Matter Intake

It has long been supposed that physical compression of the rumen from the growing fetus and uterus is involved in depression of feed intake. Negative relationships between the volume of abdominal fat and volume of rumen contents have been reported. However, the decline in VDMI in late pregnancy is even greater for cows fed diets high in concentrates than for those fed diets with more fiber. This argues that the depression in feed intake involves more that simply compression. Furthermore, parturition results in the removal of as much as 70 kg of tissues and fluids. If the purely physical constraints were involved, a rapid increase in VDMI would be expected, but this is not the case. This is especially evident when considered against the nutrient needs for increasing milk production in early lactation.

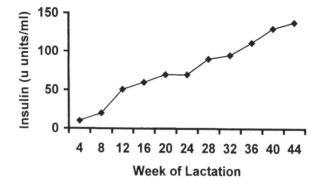

A

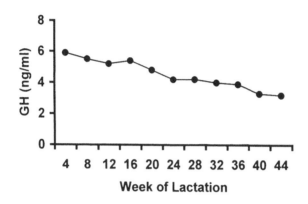

B

Fig. 9.3   Basal concentrations of insulin (**A**) and GH (**B**) in serum of cows during lactation. Data adapted from Koprowski and Tucker, 1973.

The rate of increase in VDMI is nonetheless affected by the quality and density of the diet as well as specific metabolic influences. For example, the absorption or presentation of specific nutrients to the small intestine (postruminal effects) can impact feeding behavior and intake throughout lactation. Abomasal infusion of fats reduced VDMI in many trials, but the particular response might have depended on the amount of fat, degree of saturation, or specific fatty acids. Since secretion of gastrointestinal hormones is affected by the presence of specific nutrients, osmolarity, or pH changes in the duodenum, a variety of responses is not surprising.

Voluntary DMI in most animals as well as ruminants is greatly impacted by body reserves. For example, Bines and Morant (1983) studied the effect of body condition on feed intake of nonlactating cows. They fed cows to become fat or thin and compared their ad libitum intake of straw, hay, and concentrates. There was no difference in straw consumption, but intakes of hay and concentrates were

Table 9.2 Changes in serum concentrations of presumptive homeorrhetic and homeostatic hormones, tissue sensitivity, and responsiveness in selected tissues during pregnancy and lactation

	Midpregnancy	*Late pregnancy*	*Early lactation*
Homeorrhetic hormones			
Progesterone	↑	↓	↓
Placental lactogen		↑	↓
Estrogens		↑	↓
Prl	—	—	↑
GH			↑
Leptin	?	?	?
Homeostatic hormones			
Insulin		↑	↓
Glucagon	—	—	—
CCK[a] and somatostatin	?	?	?
Tissue Sensitivity			
Insulin	↑	↓	↓
Catecholamines		↑	↑
Tissue responsiveness			
Insulin		↓	↓
Catecholamines	↓	↑	↑
Liver			
Glucogneogenesis			↑
Ketogenesis			↑
Adipose tissue			
Lipogenesis	↑	↓	↓
FA esterification	↑	↓	↓
Lipolysis		↑	↑
Glucose utilization		↓	↓
Skeletal muscle			
Protein synthesis		↓	↓
Protein degradation		↑	↑
Glucose utilization		↓	↓

Source. Data adapted from Ingvartsen and Andersen, 2000.
Note. ↑ indicates increase, ↓ indicates decrease, — indicates no change, and ? indicates it is unknown for ruminants. It should also be understood that not all ruminants are necessarily equivalent (e.g., there are differences in placental lactogen in cows vs. sheep).
[a]CCK = cholecystokinin.

reduced about 23 percent in overconditioned cows. They proposed that the more rapid catabolism of lipogenic substrates in thin cows minimized the excessive accumulation of specific metabolites in the blood and that this acted to reduce the concentration, an inhibitor(s) of appetite. Regulation of intake is especially complex in ruminants because of the impact of rumen fermentation of dietary carbohydrates and biohydrogenation of lipids. However, studies with nonruminants clearly illustrate the effects of various hormones and metabolites on VDMI.

Some especially interesting data have come from parabiotic experiments. Parabiosis is the union of two living individuals by sharing a common blood supply. Experimental surgical techniques have been used to study the involvement of metabolites and hormones on regulation of various physiological systems and especially feed intake in these animals. In early experiments to determine if humoral signals were involved in coordinating BW and intake, Hervey (1959) joined a pair of rats, one of which had a lesion of the ventromedial nucleus of the hypothalamus. Individuals with lesions of this so-called satiety center express hypophagia and dramatically lose weight. When the animals were joined, the normal animal also exhibited hypophagia and weight loss. Subsequent studies (Parameswaran et al., 1977) showed that stimulation of the lateral hypothalamus of one parabiotic rat induced a rapid increase in feeding behavior and intake and dramatic weight gain in the stimulated animal but its normal companion showed reduced intake so that the animal lost weight and progressively starved itself to death. These experiments clearly indicate that a circulating factor(s) was acting to regulate VDMI. Subsequent studies with genetically obese animals showed that that obesity can occur when there is a deficiency in the adipose-derived satiety factor leptin or when there is a defect in its receptor. However, it is unclear if more acute feeding and satiety responses associated with meals involve leptin or perhaps leptin in combination with other regulators. Table 9.3, from the extensive review by Ingvartsen and Andersen (2000), provides a list of hormones and peptides known to alter feed intake. In addition, a variety of nutrients and metabolites impact feeding behavior either directly or by alterations in secretion of hormones and growth factors.

Role of the Hormone Leptin

As discussed above and summarized in recent reviews (Ahima and Flier, 2000; Ingvartsen and Andersen, 2000), evidence for an adipose tissue–derived homeostatic regulator of feed intake has accumulated for a number of years. These studies built on the proposals by Kennedy (1953), stating that the amount of energy stored in adipose tissue mass represents a steady state between energy needs and energy derived from feed intake. Since adipose tissue tends to be relatively stable for long periods in many mammals, he suggested that there must be a regulatory mechanism that effectively monitors changes in energy stores to elicit the needed change in feed intake to "restock" adipose reserves when demand is higher but to conversely reduce "deliveries" during periods of lower energy demand.

Table 9.3 Hormones and peptides known to increase or decrease voluntary feed intake

Increase intake	Decrease intake	
β-Endorphin	Agouti-related peptide	Insulin
Dynorphin	Anorectin	Leptin
Galanin	Bombesin	Neurotensin
GH-releasing hormone	CCK 8 and 33	Norepinephrine (β)
Neuropeptide Y	Corticotropin-releasing hormone	Oxytocin
Norepinephrine (α2 receptors)	Cyclo-His-Pro	Satietin
Melanin-concentrating hormone	Dopamine	Serotonin
Melanocyte-stimulating hormone	Enterostatin	Somatostatin
Opioids	Estrogen	Substance P
Orexin A and B	Gastrin-releasing peptide	Thyrotropin-releasing hormone
Progesterone	Glucagon	Vasopressin
Peptide YY	Glucagon-like peptide-1	Xenin

Source. Adapted from Ingvartsen and Andersen, 2000.
Note. CCK = cholecystokinin.

This concept of a circulating fat-derived regulator of feeding behavior was greatly supported by the discovery of genetic mutations in mice, obese (ob) and diabetes (db) phenotypes. Both of these recessive mutations lead to hyperphagia, decreased activity, and early onset of obesity. Parabiosis of wild-type mice with ob/ob mice suppressed the weight gain in the defective mice, but parabiosis of wild-type mice with db/db mice caused marked hyperphagia and weight gain in the normal mice. This discovery led to the idea that the ob gene locus was essential for the production of a circulating satiety factor and that the db locus encoded for a molecule capable of responding to this circulating agent. The product of the ob gene was subsequently named *leptin* (from the Greek word *leptos* for thin) because of the effects of the protein to reduce feed intake and body weight when injected into leptin-deprived or normal animals. Leptin satisfied many of the requirements of the adipose tissue regulator envisioned by Kennedy many years ago. Specifically, leptin is proposed to prevent obesity by reducing feed intake and increasing thermogenesis by affecting the hypothalamus. These initial reports stimulated tremendous interest in leptin as an obesity preventative or weight control agent in humans. However, like many aspects of homeostatic or homeorrhetic regulation, simple answers are not often sufficient. Although leptin clearly can provide a signaling pathway between adipose tissue and the central nervous system for monitoring of adipose tissue stores, the wide distribution of leptin receptors as reported in many studies indicate that leptin affects many tissues and physiological systems.

Leptin is primarily produced in adipose cells and in nonruminants circulates both in a free form and bound to other proteins in circulation. Energy stores influence expression of the leptin gene as shown by increased adipose tissue leptin mRNA and serum concentrations in obese mammals. There is also a positive correlation between body fat stores and leptin concentrations in blood. Secretion occurs with a circadian rhythm and may show pulsatile secretion, as do many other hormones. Although adipose tissue is the major source of leptin, relatively lower levels of expression are found in many other tissues. It may be that local tissue production of leptin is also important in addition to effects mediated by changes in circulating concentrations.

Cloning studies of the leptin receptor (Ob-R) indicate the presence of at least six leptin receptor isoforms that are derived by alternative splice variants of the mRNA coding for the receptor. The receptor belongs to the family of cytokine receptors, which includes receptors for interleukins, granulocyte colony–stimulating factor, and Prl. Each of the leptin isoforms has identical extracellular ligand-binding domains, but they differ at the carboxy terminal end of the molecule or the cytoplasmic portion. Most of these variants have short intracellular domains of 0–40 amino acids, collectively indicated as the short or Ob-R_S receptors. The single long form of the receptor (Ob-R_L) has an intracellular domain of 303 amino acids. Differences among the isoforms mean that there can be a great deal of variation in the signaling cascade stimulated by the binding of leptin to a particular receptor. Since expression of receptor isoforms is not uniform among target tissues, this adds an additional layer of complexity to understanding the physiological effects of leptin stimulation. For example, binding of leptin to the Ob-R_L receptor or to one of the short forms of the receptor (Ob-R_a), with a 34 amino acid cytoplasmic tail, activates the tyrosine kinase Jak2, IRS-I, and MAP kinase pathway. However, only by binding to the Ob-R_L receptor can leptin also activate the Stat-3 signaling pathway. It is known that stimulation of Stat-3 is essential for many of the usual actions of leptin since the obesity defect in mutant db/db mice is caused by a failure of these mice to express the Ob-R_L form of the leptin receptor. In normal animals the Ob-R_L receptor is expressed at the high levels in regions of the hypothalamus involved in regulation of feed intake. This may mean that other forms of the leptin receptor are involved in physiological responses not directly related to nervous system control of feed intake. Unfortunately, studies in domestic animals, and especially dairy animals, are very limited (Houseknecht et al., 1998). However, fasting increases the expression of the Ob-R_L receptor in the sheep hypothalamus (Dyer et al., 1997), and rather interestingly, the sheep mammary gland expresses both the Ob-R_L and Ob-R_S leptin receptors, with the greatest expression in the epithelial cells (Laud et al., 1999). There are also preliminary reports that leptin stimulates the growth of bovine mammary epithelial cells in culture.

If extrapolated from studies with other species, it would be expected that blood levels of leptin would increase in dairy cows during the latter portion of lactation as the animals return to a positive energy balance status. With the onset of parturition and demand for energy to support lactation, leptin concentrations would be expected to decrease. If this occurs, the usual expectation would be an increase in feed intake, but regardless of anticipated changes in circulating leptin at this time, VDMI is relatively slow to recover. This may mean that leptin concentrations do not behave as expected in dairy cows at this time or more likely that control of VDMI is more complex than anticipated. The recent development of a radioimmunoassay specific for bovine leptin (Ehrhardt et al., 2000) should aid further studies to better understand relationships between leptin and physiological adjustments of support lactation in dairy cows.

Other Hormones and VDMI in Ruminants

Short-term effects of insulin on satiety in ruminants is believed to involve relationships between appearance of short-chain fatty acids and induction of insulin secretion since rumen fermentation of dietary carbohydrates minimizes effects of meals on blood glucose. Insulin can act as a modulator of neural activity as shown many years ago by its ability to reduce voluntary food intake if injected into the ventromedial nucleus of the hypothalamus of rats. Since this time, it has been learned that circulating insulin is transported across the blood-brain barrier. Insulin binds to receptors on brain endothelial cells and is transferred by transcytosis into the brain tissue. Once freed, there are specific insulin receptors on many neurons and glial cells, but the concentration of receptors is highest in regions of the hypothalamus associated with feeding behavior. The effects of insulin on the brain are believed to be chronic since injection of low levels of insulin into the ventricles reduce intake and body weight in rats. In support of this intraventricular treatments with antibodies to insulin increase intake and improve weight gains. It seems likely that insulin plays a role over the long term in control of intake and body condition in ruminants. However, since insulin concentrations are suppressed in early lactation in dairy cows, it seems unlikely that insulin is directly responsible for the depressed intake of cows in late gestation and early lactation.

The pancreatic islets of Langerhans and endocrine cells of the intestinal tract also produce glucagons. Produced as a prohormone, the processing of the protein can lead to production of several fragments with varying biological activities. In the mucosal cells of small intestine, processing of proglucagon yields glicentin, GLP-1, GLP-2, and other fragments. Glucagon and GLP-1 are likely the most important of these in control of feeding behavior. Either systemic or intraventricular injection of glucagon reduces feed intake in rats, but the liver likely mediates the effect since the hepatic vagotomy prevents this response. In addition, infusion of glucagon into the hepatic portal vein is more effective than infusion in the vena cava. Last, treatment of rats with antibodies against glucagon prior to feeding causes an increase in the period of feeding and an increase in meal size.

GLP-1 is widely produced, and its structure is conserved across many species. This supports the idea that GLP-1 is an important regulator in many mammals. For example, injections of GLP-1 markedly inhibit feeding in fasted rats. Its site of action is likely in the paraventricular nucleus of the hypothalamus and the amygdala. The GLP-1 is believed to pass the blood-brain barrier. Limited studies suggest glucagon and GLP-1 are also important in ruminants, as shown by the capacity of glucagon to reduce feed intake in sheep. Interestingly, concentrations of GLP-1 are higher in lactating compared with nonlactating sheep, but there is no direct evidence to suggest changes in GLP-1 are involved in the periparturient reduction in VDMI in dairy cows.

Studies of the nutritional actions of cholecystokinin (CCK) have a long history. It is the best studied of the endogenous hormonal satiety signals. Its name comes from its ability to cause contraction of the gall bladder to secrete bile, especially following feeding of a meal rich in fats. It is primarily produced by enterocytes of the duodenum and jejunum, but sensitive immunocytochemical and in situ hybridization assays show that CCK and mRNA for CCK are produced in many areas of the intestinal tract as well as the central nervous system. CCK is increased in the blood of cows by the delivery of fats to the small intestine (Choi and Palmquist, 1996). This may be especially important with increased formulation of "nontraditional" diets, which contain additives designed to minimize processing of lipids before arrival in the intestinal lumen as well as more general attempts to supply energy for increased milk production by feeding more fat.

The receptors for CCK in peripheral tissues (CCK-A) and the brain (CCK-B) are different in structure. This suggests there may be multiple pathways by which secretion of CCK might impact satiety in addition to effects mediated by binding of CCK to CCK-B receptors in the brain. It is has long been known that stimulation of the vagus nerve is related to satiety. For example, binding of CCK to vagal CCK-A receptors is believed to promote passage of sensory impulses to the ventromedial and paraventricular nuclei of the hypothalamus. Effects in ruminants are less well defined, but injections of antibodies to CCK into the ventricles of the cerebrum cause hyperphagia in sheep.

Originally isolated and characterized as the hypothalamic-releasing hormone, which acts to inhibit secretion of GH, it is now known that somatostatin is found in multiple sites in the body and has many effects unrelated to control of secretion of GH. Experiments that suppress synthesis of somatostatin in the central nervous system often also result in suppression of feed intake. Some experiments have shown that peripheral administration of somatostatin can also suppress feed intake via a vagus nerve–mediated pathway (similar to CCK). This supports the notion that local intestinal tract secretion of somatostatin after feeding may also be involved in regulation of feed intake. Results from immunization of cattle against somatostatin indicate that changes in circulating somatostatin also affect VDMI. For example, in a metastatistical analysis of 11 independent studies, Ingvartsen and Sejrsen (1995) reported that immunization against somatostatin in growing cattle produced an average increase in VDMI of 4.2 percent, an 11.4 percent higher average daily gain, and a 3.3 percent improvement in feed conversion ratio.

As with hormones and peptides, there is a staggering list of nutrients and metabolites that may play a role in satiety responses associated with meal feeding. However, longer-term changes during the periparturient period have a dramatic effect on the capacity of the cow to support subsequent milk production. As illustrated by the rise in NEFA (Fig. 9.2) in late pregnancy, large amounts of adipose tissue are mobilized at this time. There is also a negative correlation between VDMI and NEFA concentrations in this period. It is speculated that changes in levels of NEFA, glycerol, and/or ketones are directly involved in this loss of appetite. Regardless, the nutrient management of the cow in late pregnancy also impacts intake not just at the time of calving—responses carryover into lactation. There is a strong negative correlation between body condition before calving and VDMI in early lactation. Overconditioned cows are generally more susceptible to metabolic disorders in early lactation, and the problems are exacerbated by a slower recovery of VDMI.

Ketosis and Other Metabolic Disorders

Although it is usual to consider metabolic problems of the lactating cows in isolation, as older reviews suggest (Littledike et al. 1981; Kronfeld, 1982), complex physiological interactions produce the classic clinical signs attributed to most of these disorders. Likely the most common metabolic disorder of the lactating cow is ketosis. Ketosis can be simply defined as the accumulation of excess concentrations of acetoacetic acid (AAA), β-hydroxybutyric acid (BHBA), and the decarboxylation products acetone and isopropanol in various body fluids. It is most likely to occur in dairy cows in early lactation (between 2 and 6 weeks postpartum). Symptoms can include decreased appetite, lethargy, decreased milk production, reduced body weight, and an acetone-like odor of milk or breath. However, the disorder can be either subclinical or clinical. With clinical ketosis, the need for treatment and losses in milk production become readily apparent, but subclinical ketosis is much more problematic. In the absence of overt testing, these cows are often described as "not doing as well as expected." Kronfeld (1982) distinguished four classifications of ketosis:

1. Primary underfeeding ketosis—this is essentially a result of poor management (i.e., failure to offer enough acceptable feed to the cow).

2. Secondary underfeeding ketosis—the cow's VDMI is reduced by disease.

3. Ketogenic ketosis—the cow is consuming a diet with elements that promote production of ketones.

4. Spontaneous ketosis—the cow is consuming an adequately balanced diet, but ketosis occurs nonetheless.

Whatever the cause, lactation ketosis is a worldwide problem and is seemingly most prominent in high-producing herds. However, its incidence, especially the incidence of subclinical ketosis, can vary substantially between herds irrespective of average milk production. This suggests that its etiology is complex. Some common features are that ketotic cows are usually in negative energy balance and that frequency of clinical ketosis is often greatest at about the time of peak milk production postpartum. Two reliable biochemical changes are a reduction in blood glucose and increased concentrations of ketones in blood, urine, and milk. This has led to renewed interest in development of reliable screening methods to detect subclinical ketosis via monitoring of ketones, especially in milk samples. Animals destined to develop ketosis seemingly fail to maintain blood glucose concentrations so that the energy demands begin to be met by inappropriate overmobilization of adipose stores. Increased catabolism of the fat leads to elevations in blood lipids and transport of fatty acids into the liver in greater quantities than the liver can metabolize. This can then cause development of lipid deposits in the liver tissue (fatty liver disease), impairment of liver function, and overproduction of ketones. Since the capacity of nonhepatic tissues to metabolize ketones is limited, excess ketones accumulate in blood and subsequently appear in milk and urine. Acute treatments typically involve glucose infusions or injections to provide alternative energy substrates and/or treatment with glucocorticoids to stimulate the cow's own capacity for gluconeogenesis. Reasons why some animals seemingly readily adapt to make the metabolic adjustments required for onset of lactation and high milk production are unknown. It is interesting, however, that genetically superior animals (with respect to milk production) often secrete more GH and that one of the salient properties of GH is to promote mobilization of nutrients.

Paradoxically, excessive overfeeding of cows during the dry period, which would logically allow the accumulation of adipose tissue stores for subsequent use, actually impairs the capacity of the cow to mobilize tissue nutrients in early lactation. Most nutritionists consequently recommend that cows be moderately fed during late lactation and that concentrate feeding be increased only just before calving and then into early lactation. Prevention of ketosis is focused on management of feeding practices in the dry period and in early lactation. Since overfeeding and excessive weight gain in the dry period adversely affect the capacity of the cow to mobilize nutrients, attention to dry cow management is essential. Because of the economic problems associated with ketosis and the subtle nature of subclinical cases, there has been increased attention directed toward development of easy-to-use cow-side tests. Blood concentrations of BHBA greater than 1200 μmol/L can be used to classify normal from subclinically ketotic cows. However, blood sampling is not a routine practice in most dairies so that semi-quantitative cow-side tests for assay of milk ketones have been developed and tested with promising results (Enjalbert et al., 2001; Geishauser et al., 2000).

Parturient Hypocalcemia (Milk Fever)

Hypocalcemia, or milk fever, involves a metabolic maladjustment in early lactation in which the demands of the mammary gland for calcium needed for milk production can produce a potentially life-threatening reduction in circulating calcium. Because of the need to maintain blood calcium within relatively narrow boundaries, and widespread physiological demand for calcium (e.g., muscle contraction, cell signaling), symptoms can vary. However, the usual symptoms include low body temperature, inability to stand or unsteadiness, and reduced feed intake. Incidence is highest in Jersey cows, but within breeds it is usually associated with high milk production. Typical changes in blood metabolite profile include reduced calcium, phosphorous, and parathyroid hormone but increased calcitonin. It is clear that the capacity of the cow to mobilize calcium from bone and to maximize absorption of calcium from the diet is especially important with the onset of lactation.

A common treatment is the intravenous infusion of a calcium gluconate solution. However, it has been repeatedly demonstrated that overfeeding of calcium during the dry period increases susceptibility to the disease. Thus, careful attention to dry cow management and feeding is critical toward efforts to minimize incidence.

Acidosis and Laminitis

Because of the marked nutrient and energy demands of lactation in high-producing cows, it is tempting to increase dietary energy by increasing carbohydrate or concentrate feeding. However, it is important not to impair normal rumen functioning in attempts to maximize energy intake. Bovine lactic acidosis syndrome results from large increases in lactic acid in the rumen, which are usually produced by feeding diets that are too high in ruminally available carbohydrates or feeding forages that are too low in effective fiber. Clinical problems can range from simple reductions in VDMI to death. Other physiological consequences of acidosis can include laminitis.

As reviewed (Nocek, 1997), acute acidosis is characterized by a reduction in ruminal pH (<5.0), a large increase in lactic acid and volatile fatty acid concentration, and a decrease in rumen protozoa. If this is not corrected, these conditions promote dramatic changes in the microflora of the rumen (e.g., increased growth of *Streptococcus bovis*) and other microorganisms so that those that synthesize lactic acid begin to outnumber those that use lactic acid as an energy source. As the pH continues to decline, the population of *Streptococcus bovis* also declines, but various lactobacilli fill this niche in the rumen ecosystem, and the pH drops even further. These conditions reduce rumen motility and can promote ruminitis and hyperkatosis, both of which directly impact the health of the epithelial lining of the rumen. Rumen acidosis in turn can lead to metabolic acidosis and a variety of systemic problems.

It has also been suggested that subclinical acidosis may actually be a greater problem in many herds simply because acute clinical cases are more likely to produce specific clinical signs so that the animals are treated. In subclinical cases more vague symptoms, such as cows being slightly off feed, decreased feed efficiency, or lower than expected milk production can easily be ignored or attributed to other problems, feed quality, or feeding management so that acidosis is never suspected.

Nocek (1997) described links between chronic subclinical acidosis and laminitis as progressing through four phases. In the first, the activation phase, low rumen pH progresses to metabolic acidosis and a reduction in systemic pH. This is believed to produce a vasomotor response in the digits that ultimately causes abnormal arteriovenous shunts, increased local congestion and pressure, edema, hemorrhage, and thrombosis of the solar corium of the hoof. A second phase can lead to mechanical damage, local ischemia, and local tissue death. In phase three, loss of nutrients to the epidermis can lead to degeneration of the stratum germinativum and failure of sufficient cell replacement. In the fourth phase separation between the strata germinativum and the corium of the hoof appears. This can cause a breakdown between the dorsal and lateral laminar support of the hoof tissue. The laminar layer can separate, and the pedal bone shift from its normal position, causing compression of the soft tissue between the bone and the sole and consequently cause a variety of foot problems responsible for altered gate, pain, and difficulty in locomotion.

Many interacting factors are likely involved in the etiology of laminitis, but nutrition is considered critical in understanding onset and progression of the disease as well as development of management techniques to minimize occurrence. As with many metabolic diseases of dairy cows, the transition period around calving is especially important. Lameness is most common in the first 120 days of lactation and is more likely in older cows. Subclinical forms of the disease are especially costly because of lost milk production and poor animal performance blamed on other ills. Although carbohydrate nutrition is especially important, attention to stall comfort, sufficient stall space, and exercise are recognized contributing factors.

Factors Affecting Milk Production

While metabolic diseases and mastitis are important variables in regulation of milk production in a dairy herd, many other elements can affect milk production and can be addressed by changes in herd management. In a broad sense *environment* refers to all factors other than genotype that influence the cow's milk production and health. These environmental factors can affect milk production in rather apparent ways (e.g., lost milk production caused by reduced appetite and VDMI during periods of high heat load). Other problems (e.g., stress related to milking management or poor housing conditions) may be less apparent. Regardless, the complex interactions and interplay between physiological systems to support high levels of milk production mean that highly successful managers must be aware of conditions that maximize animal health and comfort if for no other reason than cows treated in this way are also likely to be the most productive and profitable.

Environmental Conditions

The thermoneutral zone is the temperature range to which the cow can be exposed without an impact on basal metabolic rate so that the cow maintains normal body temperature. For the typical dairy cow, this is between 10 and 20°C. Both heat and cold stress increase maintenance requirements, but in cold stress feed intake typically increases so that milk yields are usually not impaired until temperatures are below –5°C. Production losses with heat stress are more apparent and severe because appetite is typically depressed, water intake increased, and intake of needed nutrients proportionally reduced because of increased metabolic maintenance demands. Figure 9.4 illustrates some of the effects of ambient temperature on the maintenance requirements and milk production of the dairy cow.

Given the effects of temperature on lactational performance, it is useful to review factors affecting temperature regulation. Ambient temperature, airflow, relative humidity, shading or cover, and animal health influence heat loss. Heat gain is impacted by temperature, solar radiation, vegetation, and location. It is readily apparent that latitude has a major effect on both heat loss and heat gain. Latitudes between 30° north and 30° south are tropical or subtropical, yet 50 percent of the world's cattle are in these locations. Thus, problems associated with heat stress are especially important in these areas. Latitudes between 30 and 60° north and south are considered temperate, but problems with heat or cold stress can plague animals in these regions at certain times of the year. For example, environmental temperature will increase the respiratory rate (the primary mechanism whereby European-evolved breeds of dairy cattle dissipate heat) approximately five-fold as the temperature increases from 20 to 40°C. To put this in perspective, inherent heat production in lactating cows is about twice that of nonlactating cows. Heat is lost by four mechanisms: (1) convection, (2) conduction, (3) radiation, and (4) evaporation. The first three routes are effective along a temperature gradient. Thus, changes in air temperature or solar radiation easily affect heat exchange. In contrast, evaporation is closely tied to the vapor pressure gradient so that relative humidity has a major impact.

Just as individuals vary in their responses and preferences for surrounding environmental conditions, animal factors also affect the degree of apparent response and stress associated with specific conditions. For example, young animals and neonates with less insulation and higher surface area to mass ratios gain or lose heat more rapidly than older animals. This explains the often devastating effect of cold damp conditions on survival of newborn calves or lambs compared with heifers or cows. Lactating animals are more cold resistant but more heat sensitive because of the increased feed intake and heat production during lactation. Certain breeds of cattle are also better adapted to either hot or cold environments. *Bos indicus* cattle, which evolved in the tropics, are more tolerant of heat stress than *Bos taurus* cattle, which evolved in temperate climates. But within this latter group, Jersey cattle are better adapted to living in the tropics than Holstein cattle.

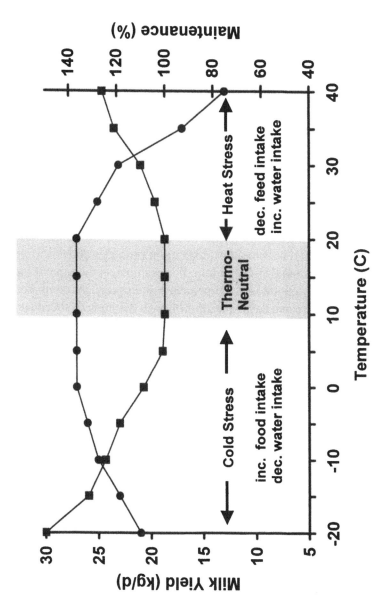

Fig. 9.4 Effect of ambient temperature on maintenance requirements (●) and milk yields (■) in cows. Maintenance requirement is expressed as a percentage of maintenance at 20°C. Adapted from Collier, 1985.

Animal Factors Affecting Milk Production

Milk yields increase as cows mature until about 8 years of age, but the change is most pronounced in the first few years. After this time production progressively declines. For example, mature cows produce about 25 percent more milk than 2-year-old heifers. Best estimates suggest that increased body weight accounts for about 20 percent of increased production and about 80 percent from increased udder development during repetitive lactation cycles. Large cows generally produce more milk than small cows, but as indicated in Chapter 1, milk yield does not vary in direct proportion to body weight but more closely corresponds with the 0.7 power of body weight, which estimates the surface area of the cow.

It should be no surprise that physiological changes associated with reproduction can also affect milk yield. Cows in estrus frequently exhibit a temporary decrease in milk production. In contrast, cows that develop follicular ovarian cysts can produce more milk (adjusted for days not pregnant) than reproductively normal herd mates. This supports the idea that endocrine changes associated with appearance of cystic ovaries stimulate increased milk production rather than the notion that cows with high milk production are more likely to produce follicular cysts. The longer that cows are cystic, the greater the production in comparison with normal herd mates. Similarly, anestrous cows produce more milk than cows with ovarian cysts producing nymphomania.

Onset of pregnancy also reduces milk production during concurrent lactation. In a 365-day period, a cow bred at 3 months postpartum produces about 350 kg less milk than a cow bred at 8 months postpartum. For example, at the time of the third trimester average milk production is 20 percent less in pregnant compared with nonpregnant cows for the same days in milk. Although twinning in sheep and goats is associated with increased milk production, in dairy cows twinning often leads to dystocia and decreased milk production. As previously discussed, a non-lactating or dry period of 40–60 days between lactations is required for optimal milk production in the subsequent lactation. Either shorter or longer periods will impair milk production in the next lactation.

In summary, common animal factors linked with increases in milk yield include good body condition at the time of calving, calving in fall or winter, housing in moderate or cool temperatures, increased body weight, advancing age, and an increased plane of nutrition. Animal factors likely to decrease milk production include dry period too short, high temperatures and humidity, spring or summer calving, advanced stage of gestation, diseases that affect the udder directly or reduce feed intake of the cow, or a decreased plane of nutrition. Interestingly, factors that act to increase the milk yield of the cow are also likely to reduce milk fat percentage.

For example, diets relatively high in concentrates but low in roughage and fiber routinely reduce milk fat percentage. The exact mechanism (or mechanisms) of the milk fat depression at the level of the mammary epithelial cell is obscure, but it is accompanied by changes in rumen fermentation. Rumen acetate production is usually reduced, but there is an increase in the rumen propionate production and a decrease in the rumen pH. The feeding of sodium or potassium bicarbonate, magnesium bicarbonate, magnesium oxide, and calcium hydroxide, in part, prevents the depression in the milk composition caused by feeding a restricted roughage ration. A general rule is that diets that increase total milk production usually reduce milk fat percentage but not necessarily total fat yields.

Role of Stress on Milk Production and Health

Stress in a general sense is a normal condition of everyday life for all animals, including dairy cows. Stress-inducing agents, called *stressors,* can come from physical environmental conditions or from social conditions within herds of cattle. Whatever the source, these stressors can disrupt homeostasis and ultimately produce imbalances in endocrine, immune, and neural systems that can negatively impact productivity and health. The effects on milk production of some of the apparent environmental stressors (temperature, wind flow, precipitation, solar radiation) were previously discussed. Other less apparent stressors include air quality, available space for rest or feeding, transport, and handling. Irrespective of the source, stressors cause secretion of cortisol and catecholamines, hormones typically associated with stressful responses. In the short-term, increased secretion of these hormones enhances capacity to respond to stress by promoting mobilization of energy reserves. However, prolonged secretion of cortisol can impair immune function. Indeed, clinical injections of synthetic glucocorticoids are a common treatment to reduce local inflammation. More closely tied to milk production, elevated epinephrine concentrations can reduce mammary blood flow and interfere with the action of oxytocin to reduce milk yields 50 percent or more. Altered behavior of cattle caused by crowding at the feed bunk or in free stall areas can serve as a symptom of stress caused by inadequate attention to animal management and social interactions. As skilled managers quickly learn, providing comfortable, well-ventilated areas with adequate space for feeding and rest minimizes distress and discomfort and reduces likelihood of injury as well as bacterial infection, leading to mastitis. For example, after 4 hours of restraint cows spent more time lying down, spent less time eating, and were more aggressive. The simple lesson is that reducing stressors in the dairy cows' environment will reduce blood cortisol and epinephrine, improve milk quality, decrease mastitis, and improve animal well-being (Gwazdauskas, 2002).

Sir Hans Selye, a pioneer in the understanding of the physiology of stress, unraveled basic principles in the 1950s. The components of the stress response include the initial perception of an event that disrupts homeostatic mechanisms in the body. The perception of stress activates neural pathways in the cerebrum and hypothalamus leading to release of corticotrophin-releasing hormone (CRH), which activates the anterior pituitary to secrete adrenocorticotropic hormone (ACTH), which in turn causes secretion of glucocorticoids from the adrenal cortex.

For dairy cows like other animals, the magnitude of response to a perceived stress can vary dramatically, just as some people are easily disturbed but others exhibit little reaction to the same presumptive stress. The degree of response is affected by the individual's assessment of the current environment, previous experiences, and tolerance of change. Responses to management stressors can include changes in behavior, physiology, or even anatomy. For example, chronic stress can cause increase in size of the adrenals and a decrease in size of the spleen due to lymphoid involution. With acute stress, there are increases in plasma glucocorticoids, mainly cortisol, and catecholamines. There are changes in the concentration of blood leukocytes due to long- and short-term stressors. These changes subsequently affect the type of response that results following bacterial and viral challenges to the immune system. With high levels of stress, viral diseases are more common.

Fortunately, dairy cattle can adapt to many types of environments. However, in developed countries with intensive husbandry, selection of dairy cattle for increased milk production has likely produced physiological demands that can make it difficult for these cows to easily respond to added stressors. This probably explains the appearance of more physiological problems in early lactation than at other times of the lactation cycle. Indeed, a common clinical method to evaluate stressors and their impact on productivity is to evaluate and monitor the physical and psychological components of animal well-being. Indicators of reduced animal well-being include unusual changes in behavior, especially time spent eating and resting, appearance of pathological and immunological lesions, altered physiological and biochemical characteristics associated with failing homeostasis, or poor reproductive and productive performance (Pajor et al., 2000).

Interestingly, the influence of elevated cortisol or ACTH on milk production and milk constituents during lactation is equivocal. Cows treated with low doses of corticosteroids or ACTH exhibit increases, decreases, or no change in milk yields and milk composition. Injection of high doses of ACTH for extended periods of time leads to increases in corticosteroids and glucose with subsequent decreases in milk yield during the injection period. Declines in milk protein yield suggest suppression in milk protein synthesis accompanying decreases in milk yield.

Regardless, social order or hierarchical rank is important within herds because it acts to minimize behavioral stressors. In short, dominant individuals regulate the activities of the herd. Social dominance is associated with age, body weight, and seniority, and it is related with milk production. Dominant behavior impacts the amount of time spent at the feed bunk and, thus, affects availability of nutrients in confinement housing systems. Appearance of behaviors caused by crowding depends on the density, contact, and activity of cattle. The period just after animals are milked, ~1 hour, is frequently a time of peak competition between animals. Reducing available space increases agonistic social behaviors and may be perceived as a stressor. Behavior of cattle is altered when <0.67 of free stalls and/or <0.2 m of linear feed bunk space are/is available per cow. An accepted physiological measure of stress is to measure the response of the adrenal gland to a challenge with exogenous ACTH. Corresponding with the increased percentage of time cows spend occupying free stalls when <0.67 of stalls were provided per cow, the greater competition for free stall space is reflected by an increase in serum glucocorticoid response after ACTH injection. This provides a physiological measure of the animal's ability to withstand a mild stress from crowding (Friend et al., 1979).

Relocation of dairy herds occurs when dairy farmers build or expand or other facilities are sold. Transportation stress occurs with movement of cattle from one farm area to another or movement from sale areas to farm areas. Managers involved with transport of cattle should focus on safety and animal comfort. Important factors include vehicle floor material, ventilation, protection, length of the trip, weather, and time required for transport. In addition to effects of restraint, isolation, or new environment stressors, loss of appetite and loss of body mass complicate the ability of animals to maintain or return to homeostasis. Body weight is lost due to dehydration and altered feeding during transit. These factors reduce rumen fermentation and populations of rumen microorganisms. It can take a week or more for appetite to return to normal so that the reduction in VDMI alone can limit productivity.

As might be predicted, if cows are moved a short distance (7 km), milk yield decreases only during the first milking. Older cows and cows in late lactation typically have larger declines in milk yield, and a negative association between the first milking milk yield and corticosteroid concentrations has been noted in the period just prior to the move. Minor changes in production measures may be related to relatively short transit and relocation stressors. Travel over longer distances should be thoroughly evaluated to reduce the stressors involved and minimize the impact on animal well-being.

Chapter 10
Manipulation of Mammary Development and Milk Production

Although bST is an important management tool to increase milk production in lactating animals, there are other methods available to increase milk yields during established lactation or to affect mammary development. For example, a number of producers now routinely milk their cows three or even four times per day to increase yields. Although the economics of more frequent milking or use of robotic milking to allow cows to be milked on demand likely varies from farm to farm, there is a seemingly continuous demand to develop techniques to increase both milk production and efficiency of production. This chapter deals with relatively recent studies designed to determine impacts of photoperiod and nutritional manipulation on mammary development and mammary gland function.

Photoperiod Effects on Dairy Cattle

As previously discussed, it has long been recognized that milk production in dairy animals is affected by season. Further, as recently reviewed (Dahl et al., 2000), manipulation of photoperiod has also been used to increase livestock productivity for many years. Examples include manipulation of photoperiod to increase egg production in chickens or to influence breeding activity in seasonal breeders such as horses and sheep. Early studies noting the positive effect of photoperiod on Prl secretion prompted a landmark investigation by Peters et al. (1978) on the effects of increased photoperiod on milk production in cows. Specifically, they showed that exposure of lactating cows to a long day photoperiod between September and April in Michigan significantly increased milk yields (2.0 kg/day) compared with cows exposed to the ambient photoperiod. This effect has been confirmed by numerous other research groups in North America and Europe, in latitudes ranging from 39 to 62° north. It is, however, important that animals experience at least a minimal dark cycle, since animals on a continuous lighting regimen appear to revert to a short-day photoperiodic response pattern.

In cattle and other mammals, the pineal gland as well as the eyes are involved in photoperiodic responses since either pinealectomy or blinding eliminates the rhythmic secretion of melatonin. It is believed that light entering the eyes stimulates retinal photoreceptor cells that ultimately produce an inhibitory signal in the pineal gland. Specifically, light stimulation causes an inhibition in the activity of N-acetyl-transferase, the rate-limiting enzyme involved in melatonin biosynthesis. This means that during exposure to light, melatonin secretion is suppressed. During the dark phase melatonin concentrations are increased because of the removal of inhibition. This suggests that physiological coding for day length depends on the duration of elevated melatonin concentrations. This coding acts to synchronize the animal's circadian rhythm and determine a relative dawn (i.e., the time when melatonin concentrations decrease). If light is perceived (and melatonin suppressed) approximately 15 hours after relative dawn, the so-called photosensitive period, the animal then senses day length related to that subjective dawn—this is the signal indicating a long day. If, however, melatonin is increased at this time because of darkness, the animal senses this as a short day. In practical terms this means that it is possible to take advantage of the photosensitive period by using a skeletal lighting scheme and avoid the costs of continuous lighting. Exposure of the animals to light between 14 and 16 hours after subjective dawn is as effective as continuous lighting. It is believed that the pattern in melatonin secretion caused by long days produces changes in galactopoietic or mammogenic hormones that are ultimately responsible for the effect on milk production (Dahl et al., 2000).

The well-known effect of a long day on Prl secretion in many species was the initial reason for studies on the effect of photoperiod on milk production in cows. The thinking at the time was that Prl was likely galactopoietic and therefore that increases in Prl during lactation would be beneficial. As reviewed in early sections, Prl is clearly galactopoietic in many species and critical for lactogenesis in cattle, but during established lactation there is no compelling evidence that increased Prl concentrations can increase milk production. For example, Plaut et al., 1987, showed that administration of exogenous Prl either before or after the time of peak milk production had no effect on milk production. It is, however, relevant that the study period lasted only 1 month. Since the effects of photoperiod on milk production do not typically become apparent until after the first month of treatment, it is possible that Prl is involved. However, the observation that temperatures below freezing block the usual increase in Prl with increased photoperiod, without affecting the increase in milk production, also suggests the increase in Prl is not critical.

As the use of exogenous bST demonstrates, GH is a potent galactopoietic hormone in dairy animals. However, much evidence shows that circulating concentrations of GH are not affected by photoperiod in cattle. For example, average daily concentration, number of secretory pulses, and metabolic clearance rate are not affected by photoperiod length. Lactating cows maintained in a controlled environmental chamber (lights on 0700 to 2300 hours, 50 percent humidity, 19°C) show relatively small (range ~2–10 ng/ml) changes in plasma GH, which produces a sinusoidal circadian rhythm with a peak at about 630 hours, an ultradian rhythm with a period of about 80 minutes, and evidence of an ultradian rhythm with a period of about 6 hours (Lefcourt et al., 1995). Collectively, there is little data to support the idea that the positive effect of increased photoperiod on milk production is mediated by changes in circulating concentrations of GH.

In contrast (as reviewed by Dahl et al., 2000) there is substantial evidence that IGF-I secretion in ruminants is impacted by season. Examples include correspondence between growth of red deer or reindeer in the spring and elevated serum IGF-I. Heifers exposed to 16 hours of light have greater circulating concentrations of IGF-I than controls. Moreover, heifers on long days exhibit increased mammary parenchymal growth (Petitclerc et al., 1984). Melatonin feeding to mimic the short day photoperiod also impairs mammary development in heifers (Sanchez-Barcelo et al., 1991). Dahl et al. (1997) showed that exposure of cows to a long day photoperiod did indeed increase milk production with a corresponding increase in circulating IGF-I concentrations. There was no difference in the groups for GH, IGFBP-2, or IGFBP-3 although, as expected, Prl was also increased. These data support a possible role for IGF-I in photoperiod-induced increased milk production, but mechanisms involved remain elusive.

Dahl et al. (2000) concluded that photoperiod manipulations can be used effectively for dairy cattle throughout development. For example, in the peripubertal period, heifers exposed to long days have a more rapid gain and greater development of mammary parenchymal tissue compared with heifers on short photoperiods. In the final 2 months of gestation in heifers and during the dry period of multiparous cows, short day photoperiods are recommended because these are associated with an enhanced response to long days during lactation. During lactation increased light exposure consistently increases milk yields. Dahl et al. (2002) viewed photoperiod manipulation as an additional management tool to enhance productivity. Because the effects on milk production of bST and milking three times per day are additive, it may be possible to combine these with a photoperiod treatment for maximal responsiveness.

Effect of Dietary Manipulation on Mammary Development

Especially during the past 20 years, a large number of studies have provided much data to indicate that feeding level, and especially energy intake in prepubertal ruminants, can markedly affect mammary development and subsequent milk production. Comprehensive reviews (Johnson, 1988; Sejrsen and Purup, 1997; Sejrsen et al., 2000) summarized data to support the idea that a critical interval exists prior to puberty during which an excessive rate of gain markedly reduces mammary development and future milk production. Moreover, after this period postpubertal and/or gestational increased rates of gain most likely enhance mammary development and subsequent milk production. Although there is some controversy as to the exact timing of this "critical interval," especially between breeds, and questions about the effects of diet composition, there can be no question of the existence of the critical interval. Most studies indicated there is no advantage obtained from freshening heifers weighing more than 615 kg regardless of their age. Therefore, if rapid rearing is to be successful, rates of gain must be closely monitored during the prepubertal period to ensure that overfattening does not occur.

Data in Table 10.1 provide a summary of results from an extensive Danish study on the effect of prepubertal average daily gain on subsequent milk production in each of three dairy breeds (Hohenboken et al., 1995). Clearly, there is a consistent pattern of lost milk production as prepubertal average daily gain (ADG) increases. These data confirm that feeding levels resulting in ADG above 600–700 g in heifers of large dairy breeds and 400 to 500 g in small dairy breeds have a negative influence on subsequent milk production.

Most early experiments studying effects of feeding level on mammary development and subsequent milk production generally maintained dietary treatments throughout the rearing period. However, most studies have shown that allometric mammary growth begins at about 2 months of age and ends about the time of puberty. Since patterns of mammary growth vary during the period prior to gestation, it is not unexpected that the effects of feeding on mammary development depend on the inherent pattern of normal growth during the period when dietary treatments are applied. Data adapted from Sejrsen et al. (2000) illustrate this idea (Fig. 10.1). Specifically, a number of experiments have shown that overfeeding during the prepubertal period impairs mammary development, but during the calfhood period or during the postpubertal interval, the same dietary treatment has no effect or even enhances mammary development.

Because increased rates of gain can be achieved with a variety of diets, it is not surprising that many workers have questioned if negative effects of high ADG dur-

Table 10.1 Effect of prepubertal rate of gain on milk production in different dairy breeds

Breed	N	ADG (g) to 320 kg	At calving age (mo.)	BW (kg)	FCM kg (relative)
Danish Jersey	41	362	29	341	5125 (100)
	44	487	26	353	4750 (93)
	44	557	23	329	4125 (80)
Danish Red	52	549	29	530	5675 (100)
	52	718	26	525	4900 (86)
	51	845	23	490	4700 (82)
Damish Friesian	53	579	29	513	5425 (100)
	53	731	26	500	5400 (99)
	55	858	23	498	4900 (90)

Source. Data adapted from Hohenboken et al., 1995.
Note: ADG = average daily gain; FCM = fat-corrected milk.

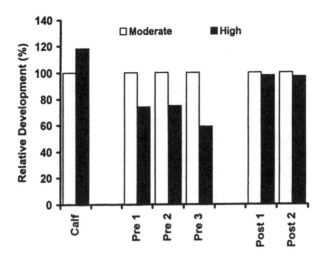

Fig. 10.1 Relative mammary growth in heifers fed at a moderate or high feeding level at different stages prior to conception for multiple experiments is illustrated. There is a consistent pattern of impaired mammary development for prepubertal heifers reared at a high rate of gain compared with moderately fed heifers. During the calf-rearing period or postpubertal period, there is no negative effect on mammary development. Adapted from Sejrsen et al., 2000.

ing the prepubertal period might not be altered depending on diet composition. Many experiments on the effects of feeding level have inconclusive effects because of variations in the concentrate to roughage ratio, so it is possible that effects attributed to differences in feeding level may be due to differences in the type of diet. Sejrsen and Foldager (1992) used 38 Red Danish heifers to investigate the effects of ration energy concentration at the same level of feeding on mammary growth and subsequent milk production. Specifically, heifers were given the same amount of energy in diets with low- (maximal barley straw added) or high-energy density (concentrate with minimal straw). Treatments were begun at 90 kg body weight and continued until slaughter or calving. Heifer ADG was 493 and 484 g/day for low-energy density (LED) slaughter and lactating animals and 498 and 475 g/day for high-energy density (HED) slaughter and lactating animals. In the slaughter group (nine and eight animals per group), parenchymal tissue mass was not different between treatments based on dissected tissue weight (412 vs. 464 g) or volume (65 vs. 59 cm^3). Neither was milk yield affected by an energy concentration ($n = 8$ and 10 per group) of 16.1 and 16.5 kg/day, respectively. These data demonstrated that mammary growth and milk production capacity are unaffected by ration energy density when heifers are raised at a low feeding level.

Capuco et al. (1995) summarized data from a similar diet composition study in which Holstein heifers between 175 kg and 325 kg of body weight were fed diets of alfalfa or corn silage to gain 725 (low) or 950 (high) g/day. Eight heifers per treatment were killed for tissue evaluations (i.e., four were killed in estrous and four during the luteal phases of their third cycle). The remaining heifers were bred during the third cycle, dietary treatments were discontinued, and all animals were fed alike in accord with normal herd practice. Actual ADG averaged 766, 792, 974, and 1026 g for low-alfalfa (LA), low-corn (LC), high-alfalfa (HA), and high-corn (HC) treatments for the slaughtered heifers. Parenchymal tissue was significantly lower in the HC diet (1056 mg) and numerically greater in the LC + LA than the HA + HC diets (1424 vs. 1957 mg). It was also found that the relative amount of epithelial tissue was reduced and adipose tissue increased in histological sections of mammary parenchyma of HC heifers compared with other treatments. The data are consistent with previous demonstrations of a deleterious effect of rapid prepubertal weight gain on mammogenesis (i.e., the HC group heifers with greatest ADG exhibited reduced mammary growth). Although the milk yields of the remaining heifers did not differ ($P > 0.05$), milk yields were numerically lower for heifers on HA and HC diets (19.7 and 19.9 kg/day; overall 19.8 kg/day fat-corrected milk [FCM]) than for those on LA and LC diets (20.6 and 21.7 kg/day; overall 21.2 kg/day FCM) (Waldo et al., 1998). This may reflect the fact that ADG among HC treatment heifers was less than in the subset of animals slaughtered (998 vs. 1026 g). It may also be that differences in ADG between low and high feeding group animals (791 vs. 994) were not sufficient to detect milk production effects in Holsteins. An additional consideration is that diets were imposed only for a portion of the critical prepubertal growth phase. Indeed, allometric mammary growth begins at about 3 months of age. Given that Danish Friesians and U.S. Holsteins are similar, it may be that detrimental effects begin to be exhibited at gains greater than 700 g/day, and thus, detection of negative effects would require comparison with heifers gaining less than 700 g/day during the critical period. It was suggested that the different effects in the heifers on the two diets could be due to the higher protein in the alfalfa diet acting to compensate for the negative effect of feeding a high-energy diet.

Kertz et al., 1987, suggested that increasing the protein content of the diet might prevent negative effects. Only a few well-controlled experiments are available that address this supposition. Data from Mantysaari et al. (1995) indicate that the protein source (urea vs. rape seed meal) did not prevent a negative effect from rapid weight gain on mammary growth in Ayrshire heifers. Heifers gained 86 kg by an average of 88 days of age at the start of the experiment. The heifers were slaughtered at approximately 220 kg at an average age of 292 and 249 for low- and high-energy diet groups. Three animals had started to cycle by the slaughter date and were killed in the luteal phase of the estrous cycle.

Not all experiments show that a high-energy diet has a negative effect in the prepubertal period on mammary development or on subsequent milk production. In some cases this may be explained by the number of animals, the relatively small differences between growth rates between treatments, or the fact that treatments were applied outside the critical period. For example, Van Amburgh et al. (1998) concluded that milk production was unaffected by the feeding level for body weight growth between 100 and 300 kg. This was evident if milk yield of the heifers was adjusted by weight at calving since high-rate-of-gain heifers weighed less at this time. Radcliff et al. (1997) also showed that a high-protein, high-energy diet increased growth rate and decreased the age of puberty in Holstein heifers but without evidence of detrimental effects on mammary development.

It is nonetheless possible that mammary growth can be affected by specific components in the diet. Lambs fed protected polyunsaturated fat between 7 and 22 weeks of age grew at a similar rate as lambs fed a high-energy diet, but mammary parenchymal weight (15 vs. 25 g) and parenchymal DNA (12 vs. 19 mg) were increased (McFadden et al., 1990). Zhang et al. (1995) reported that diets containing 15 compared with 20 percent protein had no effect on mammary growth in lambs slaughtered at 6 months of age after having been fed the diets for 100 days during the prepubertal period (i.e., beginning at weaning, 73 days of age). However, mammary glands of ewes slaughtered after 7 weeks of lactation had more parenchymal tissue (864 vs. 1156 g) and parenchymal DNA (974 vs. 1228 mg) and tended to produce more milk (1.5 vs. 1.9 kg/day).

Since it is known that bST levels in the blood are lower in rapidly growing heifers, several researchers have examined the response of such heifers to bST supplementation. Radcliff et al. (1997) evaluated the influence of elevated energy and protein in combination with daily injections of bST on growth and mammary development in 40 Holstein heifers. As expected, heifers fed for high gain weighed more and had more carcass fat. Regardless, feeding the high-energy, high-protein diet did not reduce mammary parenchymal DNA. Injections of bST increased mammary DNA by 47 percent irrespective of diet. Subsequent milk yields nonetheless were lower in animals previously given the high-protein, high-energy diet in the absence of bST treatment. However, bST given in conjunction with a high feed intake reduced age at first calving without reducing milk yields (Radcliff et al., 2000). This suggests it might be possible to use bST to negate some of the deleterious effects of overfeeding on mammary development.

Many dairy producers suggest that larger, older heifers produce more milk in their first and later lactations. However, studies of Dairy Herd Improvement Association (DHIA) records show that an average age of first calving (AOFC) of about 23 months produces maximum lifetime performance, while maximum lifetime profit is associated with an AOFC of 25 months. For example, the Pennsylvania DHIA data show little difference in first lactation yield for heifers between 23 and 28 months of age at first calving (Heinrichs, 1996). Compared with AOFC, body weight after calving is significantly more important in determining first lactation yield. Keown and Everett (1986) found that the highest yields occurred when heifers weighed between 545 and 613 kg after their first calving. Moreover, production declined when postcalving body weight exceeded 660 kg. This is probably because these heifers were overconditioned and experienced a higher than normal incidence of dystocia. Although body weight is an important measure, it is important to remember that other measures are necessary to adequately describe heifer growth and likely genetic variance associated with growth. From his review of the literature, Hoffman (1997) defined optimum body size criteria for Holstein replacement heifers at calving. Specifically, best performance was with heifers that weighed 621, 560, and 522 kg, 14 days prepartum, 7 days postpartum, and 30 days postpartum, respectively. Other aspects of growth included an average height of 140 cm at the withers, a body length of 170 cm (point of shoulder to ischium), and a body condition score of 3.5 as measured 2 weeks prior to calving.

Since within a range there is positive relationship between milk yield and body weight at calving, some producers have questioned the negative effect of overfeeding on mammary development and milk production. The relationship between prepubertal growth rates and mammary development is complicated by the fact that heifers will grow fast for one of two reasons. When heifers are fed ad libitum and grown in a good environment, feeding a diet high in energy density will result in faster body weight gain. These heifers will likely also become too fat, and with average gains greater than 1 kg/day, subsequent milk production will be decreased at least 10 percent. Within a group of heifers, however, some will grow faster than others. The heifers that gain body weight most rapidly within a group do not necessarily produce less milk once they become cows. However, the most overconditioned animals within a group are also the most likely to have the least amount of mammary parenchymal tissue. As reviewed, Sejrsen et al. (2000) suggested that selection of dairy cows for increased milk production has produced a correlated increase in growth capacity of heifers along with increased production in cows. Since heifers are usually handled in groups, this means that genetically superior animals within a group (irrespective of the feeding treatment applied) likely also grow faster. Sejrsen et al. observed that this relationship corresponded with an increase in milk yield of 0.61 kg/day or 186 kg per lactation for each 10 g increase in genetic growth capacity. Within a feeding group animals in the top 10 percent with respect to genetic growth potential would be expected to gain 100 g more per day, weigh 75 kg more at parturition, and therefore produce 6–7 kg more milk per day than the bottom 10 percent of heifers. This means that the relationship between daily gain and milk yield observed in experiments with varying feeding levels relates to the average daily gain of the entire group and not of individuals within the group. For example, heifers with the highest growth rate within a feeding group would be expected to have the highest yield. Relationships between changes in milk production due to genetic selection combined with the effect of feeding level on milk production are illustrated in Figure 10.2 (adapted from Sejrsen et al., 2000).

Although the effect of feeding programs and rate of growth in the peripubertal period on mammary development and milk production has been considered in many studies, there is still much to be learned. For example, Bar-Peled et al. (1997) showed that a higher rate of gain in the milk-rearing period has a positive effect on mammary development and function. They found that calves allowed to suckle their dams three times per day compared with calves that were limit-fed milk replacer were larger at calving and produced more milk in their first lactation. This

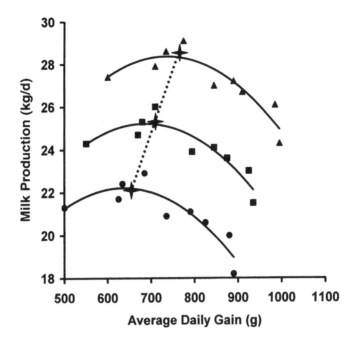

Fig. 10.2 Illustration of the expected change in optimal daily gain related to milk production for the 1980's (*closed circles*), 1990's (*closed squares*), and 2000's (*closed triangles*). The relationship between feeding level and milk yield is unchanged although average milk production has increased. The dashed line with stars represents genetic change related to milk production during this period. Adapted from Sejrsen et al., 2000.

is supported by preliminary data summarized by Sejrsen et al., 2000. Furthermore, although most rearing studies assume steady rates of average daily gain, several studies have shown that heifers managed to grow in a stair-step fashion, with periods of minimal growth followed by periods of compensatory growth, outproduce controls (Park et al., 1987).

In a study by Park et al. (1998), 40 crossbred beef heifers were placed in one of two rearing schemes when the animals reached 8 months of age. Half of the animals were reared to grow steadily, and half were fed on a schedule to provide alternating periods of feeding at 130 percent and 70 percent of National Research Council (NRC) requirements. Treatments lasted for 8 months, and the heifers were bred at 14 months and were then placed on identical management schemes. Stair-step heifers gained nearly twice as much as control in the final realimentation period. Moreover, RNA and protein contents of the mammary parenchymal tissue were higher, and the heifers were estimated to produce 6 percent more milk than controls. In a similar trial, Choi et al. (1997) fed 24 Holstein heifers in this manner on a schedule of 3, 2, 4, 2, 5, and 2 months in which the stair-step heifers were fed alternately 25 percent below and 25 percent above NRC requirements. The heifers were 6 months of age and averaged 172 kg at the start of the trial. In late pregnancy mammary tissue of stair-step–treated heifers had higher concentrations of DNA, RNA, and protein. Moreover, milk production averaged 7344 kg in stair-step compared with 6765 kg in control heifers.

In a novel dietary study, Lammers et al. (1999a, b) determined the effect of the rate of gain of heifers, coupled with treatments with estrogen implants, on mammary development and milk production. Treatments were feeding heifers so that they achieved a standard growth rate (700 g/day) or accelerated growth rate (1000 g/day) with or without estrogen implants. Treatments were initiated when the heifers were 4.5 months of age and averaged 130 kg body weight. Treatments continued for 20 weeks when the heifers were 9.5 months of age. Accelerated growth and estrogen implants both significantly reduced milk yields (7.1 and 5.1 percent, respectively). There was no significant interaction between treatments. Thus, despite apparent effects of estrogen on mammary development as indicated by changes in teat length, estrogen implants did not prevent the impairment in milk production associated with rapid rearing. Indeed, the implants reduced milk production in the standard reared heifers as well.

In summary, high milk production depends on the ability of the mammary gland to synthesize milk as well as the ability of the cow to supply the gland with necessary nutrients. The cow's capacity to provide nutrients for milk synthesis depends on size, health, and metabolic fitness during lactation. Poorly grown heifers are less competitive at the feed bunk, use significant amounts of nutrients for growth, and produce less milk (Hoffman, 1997). This means that an adequate heifer-rearing program is key to producing first-lactation animals (1) with well-developed mammary glands capable of maximum milk production and (2) with a supporting body size and body condition capable of high feed intake and the processing and delivery of nutrients to the mammary gland. There are two common mistakes in heifer management on many farms: (1) inadequate body size at calving and (2) impaired mammary development. For minimal calving problems and acceptable milk production, Holstein heifers should weigh 550 to 600 kg after calving. Unfortunately, heifers often receive little attention and grow poorly. This may be especially evident on many small and midsized farms that have inadequate numbers of heifers for optimal grouping by age. The result is that many heifers will enter the herd too small. Either faster growth or a later calving age would likely improve production and profitability. Current data suggest that milk yield may be decreased 70 kg for every 10 kg in body weight below 570 kg at first calving (Van Amburgh et al., 1998).

Management of dairy replacement heifers has a major impact on the profitability of the dairy industry. On average about 20 percent of total dairy farm expenses is spent on heifer rearing. Moreover, heifer management decisions impact the productivity, health, longevity, and profitability of the animal once she enters the milking herd (James, 2001; Van Amburgh and Tikofsky, 2001). One approach to decrease costs of raising heifers is to breed them so they calve earlier. With adequate growth, decreasing age at first calving to 24 months is usually profitable because of cost savings in facilities and labor, but some managers are now striving for calving as early as 19 to 20 months of age. If heifers are to calve as early as 19 months with adequate body size, they must average daily gains greater than 1 kg/day. This frequently occurs on farms when heifers are fed excess energy in the form of corn silage so that growth is rapid and fat deposition is high. As discussed above, mammary growth likely will be impaired, but the farmer will never know because the consequences of overfeeding in the early critical period occur long after the fact (Sejrsen et al., 2000). Impaired mammary development can occur with no change in udder size because the udder has both parenchymal tissue and extra parenchymal adipose tissue. Clearly, the lack of a noninvasive and economical method to assess mammary parenchymal development in heifers has hampered practical efforts to refine and implement the best management practices for heifers on farms.

Effect of Growth Hormone on Mammary Development

As outlined in previous chapters, the role of the pituitary gland, especially for GH in mammary ductular development, has been known in considerable detail since the 1940s. Indeed, as reviewed by Topper and Freeman, 1980, now classic rodent studies confirmed roles for Prl and GH in mammogenesis. Growth hormone and estrogen especially were deemed critical for pubertal ductal development, and synchronous secretion of estrogen and progesterone after conception regarded essential for final duct elongation and lobulo-alveolar formation. Surgical complications have made similar studies difficult for economically important ruminants. Nonetheless, fundamental aspects are certainly similar between species. For example, a study of hypophysectomized-ovariectomized goats indicates a critical role for GH in normal development (Cowie et al., 1966).

Since these pioneering studies and especially the introduction of recombinant bovine GH (bST), a number of studies have evaluated effects of exogenous bST on mammary development in ruminants. Most of these studies have not specifically tested the role of GH in normal development but have more often investigated the possibility that mammary development could be enhanced by administration of bST. Sejrsen et al., 1999, reviewed the role of GH on mammary development in farm animals. A general conclusion is that GH is important for normal mammary development and that treatment with exogenous GH can increase mammary development in heifers and at least in some cases produce small increases in subsequent milk production. The prepubertal bovine mammary gland responds to both GH and ovarian secretions. While effects of complete removal of endogenous GH on mammary development in the bovine have not been reported, a positive effect of bST on udder growth in heifers has been reported in a number of studies (Sejrsen et al., 1999). Despite demonstrations that GH treatment stimulates mammary growth before puberty, the results do not convincingly show that the effect is translated into increased milk yield. Thus, it is possible that bST may be useful as a management tool in dairy heifer husbandry, but there is certainly much to be learned to determine the exact consequences of stimulated mammary growth in the peripubertal period on subsequent lactational performance. It seems intuitive to conjecture that since impairment of prepubertal mammary growth can reduce subsequent milk yields, enhanced mammary growth at this time might improve subsequent milk production. More limited data suggest that GH treatment during late pregnancy or during the dry period might increase milk yield during the subsequent lactation. As with GH-mediated effects on milk production in lactation, mechanisms for GH stimulation of mammary development in heifers are poorly understood.

Regardless, no discussion of bST could be complete without careful consideration of the insulin-like growth factor family of related proteins. As reviewed by Akers et al., 2000, and Hovey et al., 1999, the hypothesis that the effect of GH on mammary growth is indirectly mediated via IGF-I has been the subject of increasing research. Thus, the mechanism of action of GH remains a puzzle but almost certainly involves the IGF family of related proteins, as well as locally produced factors, including receptors, binding proteins, and perhaps other growth factors. Although the liver is the major source of circulating IGF-I, the physiological importance of locally synthesized IGF-I in the mammary gland is increasingly appreciated. Local IGF-I is stroma-derived in the rodent, human, and ruminant mammary gland. IGF-I expression is also higher in the mammary stroma than in the parenchyma, and it is increased during onset of allometric mammary growth in sheep (Hovey et al., 1999). Differences between IGF-I expression in stromal and parenchymal tissue are difficult to evaluate because stromal cells surround the epithelium in the developing udder. However, IGF-I expression in mammary stroma from intact and epithelial-cleared mammary glands of sheep and cattle leaves no doubt that IGF-I synthesis occurs in the stroma (Hovey et al., 1998; Berry et al., 2001). Feeding level and bST treatment have also been shown to alter expression of IGF-I and of IGF-binding protein mRNA transcripts in the developing heifer mammary gland (Weber et al., 2000a). Specifically, Northern blot analysis of mammary parenchymal tissue from prepubertal heifers showed a major 7.5-kb IGF-I mRNA, and there was a tendency for an interaction between bST treatment and feeding level on mammary expression of the 7.5-kb IGF-I transcript. Expression of mRNA for IGFBP-3 (2.6 kb), IGFBP-2 (1.8 kb), and IGFBP-1 (2 kb) was also detected in mammary tissue. A high feeding level reduced levels of the IGFBP-1 mRNA. In contrast, expression of IGFBP-3 and IGFBP-2 mRNA was unchanged in mammary tissue by bST or feeding level. Interestingly, the capacity of mammary tissue extracts to stimulate growth of mammary test cells in culture paralleled tissue responses in the animal. That is, extracts from mammary tissue of moderately fed heifers stimulated the growth of test cells more effectively than extracts prepared from heifers fed at a high rate. Weber et al. (1999) also evaluated the effect of recombinant IGFBP-3 and IGF-I antibody addition to cultures of primary undifferentiated mammary epithelial cells when supplemented with IGF-I, serum, or mammary gland extracts from prepubertal heifers. Addition of heifer serum stimulated cell proliferation at concentrations of 4 percent and greater. Mammary extracts, on the other hand, stimulated cell proliferation much more effectively than serum or IGF-I alone, yielding a mitogenic response more that twice that induced by the highest IGF-I concentration tested. IGFBP-3 and antibodies to IGF-I were added to media containing either 50 ng/ml of IGF-I, 5 percent serum, or 5 percent mammary extracts to determine the contribution of IGF-I to mitogenic activity in serum and mammary extracts. When added with IGF-I, equimolar and greater concentrations of IGFBP-3 reduced DNA synthesis to the level of basal medium. The bioactivity of serum was reduced 73 and 43 percent by mono- and polyclonal antibodies at equimolar concentrations, respectively, while recombinant IGFBP-3 completely abolished the effect of serum. An equimolar concentration of IGFBP-3 abrogated 35 percent of the mitogenic activity in extracts. These data suggest that the IGF-I provides much of the mitogenic activity available in serum and mammary tissue, either individually or through interactions with IGFBPs.

In summary, these results strongly support the idea that effects of feeding, bST, or the ovary on mammary development in the peripubertal heifer involve the IGF-I axis of related proteins and that alterations in local tissue concentration or synthesis of these molecules is important. This does not, however, preclude the likely importance of other hormones or growth factors in control of mammary development. A prudent conclusion is to suggest that a myriad of mammogenic hormones, binding proteins, growth factors, receptors, extracellular matrix molecules, and proteolytic enzymes interact in a complex fashion to control mammary development and ultimately mammary function.

References

A

Adams, T.E., V.C. Epa, T.P.J. Garrett, and C.W. Ward. 2000. Structure and function of the type 1 insulin–like growth factor receptor. CMLS. Cell. Mol. Life Sci. 57: 1050–1093.

Ahima, R.S. and J.S. Flier. 2000. Leptin. Annu. Rev. Physiol. 62: 413–437.

Akers, R.M. 1985. Lactogenic hormones: Binding sites, mammary growth, secretory cell differentiation, and milk biosynthesis in ruminants. J. Dairy Sci. 68:501–519.

Akers, R.M. 1991. Lactation physiology: A ruminant animal perspective. Protoplasma 159: 96–111.

Akers, R.M. 2000. Selection for milk production from a lactation biology viewpoint. J. Dairy Sci. 83: 1151–1158.

Akers, R.M. and C.W. Heald. 1978. Effect of removal of prepartum secretions on secretory cell differentiation in the bovine mammary gland. J. Ultrastruct. Res. 63: 316–322.

Akers, R.M. and R.M. Kaplan. 1989. Role of milk secretion in transport of prolactin from blood into milk. Horm. Metabol. Res. 21: 362–365.

Akers, R.M. and A.M. Lefcourt. 1982. Teat stimulation-induced prolactin release in non-pregnant and pregnant Holstein heifers. J. Endocrinol. 96: 433–442.

Akers, R.M. and S.C. Nickerson. 1983. Effect of prepartum blockade of microtubule formation on milk production and biochemical differentiation of the mammary epithelium in Holstein heifers. Int. J. Biochem. 15: 771–775.

Akers, R.M. and K. Sejrsen. 1996. Mammary development and milk production. NARES 74: 44–63.

Akers, R.M. and W. Thompson. 1987. Effect of induced leucocyte migration on mammary cell morphology and milk component biosynthesis. J. Dairy Sci. 70: 1685–1695.

Akers, R.M., C.W. Heald, T.L. Bibb, and M.L. McGilliard. 1977. Effect of prepartum milk removal on quantitative morphology of bovine lactogenesis. J. Dairy Sci. 60: 1273–1282.

Akers, R.M., D.E. Bauman, A.V. Capuco, G.T. Goodman, and H.A. Tucker. 1981a. Prolactin regulation of milk secretion and biochemical differentiation of mammary epithelial cells in periparturient cows. Endocrinology 109: 23–30.

Akers, R.M., D.E. Bauman, G.T. Goodman, A.V. Capuco, and H.A. Tucker. 1981b. Prolactin regulation of cytological differentiation of mammary epithelial cells in periparturient cows. Endocrinology 109: 31–40.

Akers, R.M., W.E. Beal, T.B. McFadden, and A.V. Capuco. 1990. Morphometric analysis of involuting bovine mammary tissue after 21 or 42 days of non-suckling. J. Anim. Sci. 68: 3604–3613.

Akers, R.M., T.B. McFadden, S. Purup, M. Vestergaard, K. Sejrsen, and A.V. Capuco. 2000. Local IGF-I axis in peripubertal ruminant mammary development. J. Mammary Gland Biol. Neoplasia 5: 43–51.

Asimov, G.J. and N.K. Krouze. 1937. The lactogenic preparations from the anterior pituitary and the increase of milk yield in cows. J. Dairy Sci. 20: 289–306.

Ayares, D.L. 2000. Transgenic protein production: Achievements using microinjection technology and the promise of nuclear transfer. J. Anim. Sci. 78(Suppl. 3): 8–18.

B

Ball, S., K. Polson, J. Emeny, W. Eyestone, and R.M. Akers. 2000. Induced lactation in prepubertal Holstein heifers. J. Dairy Sci. 83: 2459–2463.

Barnes, M.A., G.W. Kazmer, R.M. Akers, and R.E. Pearson. 1985. Influence of selection for milk yield on endogenous hormones and metabolites in Holstein heifers and cows. J. Anim. Sci. 60: 271–284.

Bar–Peled, U., E. Maltz, I. Bruckental, Y. Folman, Y. Kali, H. Gacitua, A.R. Lehrer, C.H. Knight, B. Robinzon, H. Voet, and H. Tagari. 1995. Relationship between frequent milking or sucking in early lactation and milk production of high producing dairy cows. J. Dairy Sci. 78: 2726–2736.

Bar–Peled, U., B. Robinzon, E. Maltz, H. Tagari, Y. Folman, I. Bruckental, H. Voet, H. Gacitua, and A.R. Lehner. 1997. Increased weight gain and effects on production parameters of Holstein calves that were allowed to suckle from birth to six weeks of age. J. Dairy Sci. 80: 2523–2528.

Barrington, G.M., T.E. Besser, C.C. Gay, W.C. Davis, J.J. Reaves, T.B. McFadden, and R.M. Akers. 1999. Regulation of the immunoglobulin G1 receptor: Effects of prolactin on in vivo expression of the bovine mammary immunoglobulin G1 receptor. J. Endocrinol. 163: 25–31.

Bauman, D.E. 1999. Bovine somatotropin and lactation: From basic science to commercial application. Domest. Anim. Endocrinol. 17: 101–116.

Bauman, D.E. and W.B. Currie. 1980. Partitioning of nutrients during pregnancy and lactation: A review of mechanisms involving homeostasis and homeorhesis. J. Dairy Sci. 63: 1514–1529.

Bauman, D.E. and C.L. Davis. 1974. Biosynthesis of milk fat. In Lactation: A Comprehensive Treatise, vol. 2, edited by B.L. Larson and V.R. Smith. New York: Academic Press.

Bauman, D.E., R.W. Everett, W.L. Weiland, and R.J Collier. 1999. Production responses to bovine somatotropin in northeast dairy herds. J. Dairy Sci. 82: 2564–2573.

Baumrucker, C.R. and J.W. Blum. 1994. Effects of dietary recombinant human insulin-like growth factor-I on concentrations of hormone and growth factors in the blood of newborn calves. J. Endocrinol. 140: 15–21.

Baumrucker, C.R. and N.E. Erondu. 2000. Insulin-like growth factor (IGF) system in the bovine mammary gland and milk. J. Mammary Gland Biol. Neoplasia 5: 53–65.

Beal, W.E., D.R. Notter, and R.M. Akers. 1990. Techniques for estimation of milk yield in beef cows and relationships to calf weight gain and postpartum reproduction. J. Anim. Sci. 68: 937–943.

Bequette, B.J., F.R.C. Blackwell, and L.A. Crompton. 1998. Current concepts of amino acid and protein metabolism in the mammary gland of the lactating ruminant. J. Dairy Sci. 81: 2540–2559.

Bernaud, C., R.B. Dickson, and E.W. Thornpson. 1998. Roles of the matrix metalloproteinases in mammary gland development and cancer. Breast Cancer Res. Treat 50: 97–116.

Berry, S.D., T.B. McFadden, R.E. Pearson, and R.M. Akers. 2001. A local increase in the mammary IGF–I:IGFBP–3 ratio mediates the mammogenic effects of estrogen and growth hormone. Domest. Anim. Endocrinol. 21: 39–53.

Bines, J.A. and S.V. Morant. 1983. The effect of body condition on metabolic changes associated with intake food by the cow. Br. J. Nutr. 50: 81–89.

Bitman, J., D.L. Wood, S.A. Bright, R.H. Miller, A.V. Capuco, A. Roche, and J.W. Pankey. 1991. Lipid composition of teat canal keratin collected before and after milking from Holstein and Jersey cows. J. Dairy Sci. 74: 414–420.

Blackburn, D.G., V. Hayssen, and C.J. Murphy. 1989. The origins of lactation and evolution of milk: A review with new hypothesis. Mammal. Rev. 19: 1–26.

Blum, J.W. and H. Hammon. 1999. Endocrine and metabolic aspects in milk-fed calves. Domest. Anim. Endocrinol. 17: 219–230.

Bouchard, L., S. Blais, C. Desrosiers, X. Zhao, and P. Lacasse. 1999. Nitric oxide production during endotoxin-induced mastitis in the cow. J. Dairy Sci. 82: 2574–2581.

Bruckmaier, R.M. and J.W. Blum. 1998. Oxytocin release and milk removal in ruminants. J. Dairy Sci. 81: 939–949.

Butler, A.A., S. Yakar, I.H. Gewolb, M. Karas, Y. Okubo, D. LeRoith. 1998. Insulin-like growth factor-I receptor signal transduction: At the interface between physiology and cell biology. Comp. Biochem. Physiol. Part B 121: 19–26.

Byatt, J.C., P.J. Eppard, L. Murryakazi, R.H. Sorbet, J.J. Veenhuizen, D.F. Curran, and R.J. Collier. 1992. Stimulation of milk yield and feed intake by bovine placental lactogen in the dairy cow. J. Dairy Sci. 75: 1216–1223.

Byatt, J.C., R.H. Sorbet, P.J. Eppard, T.L. Curran, D.F. Curran, and R.J. Collier. 1997. The effect of recombinant bovine placental lactogen on induced lactation in dairy heifers. J. Dairy Sci. 80: 496–503.

C

Capuco, A.V. and R.M. Akers. 1999. Mammary involution in dairy animals. J. Mammary Gland Biol. Neoplasia 4: 137–144.

Capuco, A.V. and R.M. Akers. 2002. Galactopoiesis: Effects of hormones and growth factors. In Encyclopedia of Dairy Science. Academic Press (in press).

Capuco, A.V., P.A. Feldhoff, R.M. Akers, J.L. Wittliff, and H.A. Tucker. 1982. Progestin binding in mammary tissue of prepartum, nonlactating and postpartum, lactating cows. Steroids 40: 503–517.

Capuco, A.V., M.J. Paape, and S.C. Nickerson. 1986. In vitro study of polymorphonuclear leukocyte damage to mammary tissue of lactating cows. Am. J. Vet. Res. 47: 663–668.

Capuco, A.V., J.E. Keys, and J.J. Smith. 1989. Somatotrophin increases thyroxine-5'-monodeidonase activity in lactating mammary tissue of the cow. J. Endocrinol. 121: 205–211.

Capuco, A.V., D.L. Wood, S.A. Bright, R.H. Miller, and J. Bitman. 1990. Regeneration of teat canal keratin in lactating dairy cows. J. Dairy Sci. 73: 1745–1750.

Capuco, A.V., J.J. Smith, D.R. Waldo, and C.E. Rexroad, Jr. 1995. Influence of prepubertal dietary regimen on mammary growth of Holstein heifers. J. Dairy Sci. 78: 2709–2725.

Capuco, A.V., R.M. Akers, and J.J. Smith. 1997. Mammary growth in Holstein cows during the dry period: Quantification of nucleic acids and histology. J. Dairy Sci. 80: 477–487.

Capuco, A.V., R.M. Akers, S.E. Ellis, and D.L. Wood. 2000. Mammary growth in Holstein calves: Bromodeoxyuridine incorporation and steroid receptor localization. J. Dairy Sci. 83(Suppl. 1): 67.

Capuco, A.V., D.L. Wood, R. Baldwin, K. McLeod, and M.J. Paape. 2001. Mammary cell number, proliferation and apoptosis during the bovine lactation cycle: Relationship to milk production and effect of bST. J. Dairy Sci. 84: 2177–2178.

Chepko, G. and G.H. Smith. 1999. Mammary stem cells: Our current understanding. J. Mammary Gland Biol. Neoplasia 4: 35–52.

Choi, B.R. and D.L. Palmquist. 1996. High fat diets increase plasma cholecystokinin and pancreatic polypeptide, and decrease plasma insulin and feed intake in lactating cows. J. Nutr. 126: 2913–2919.

Choi, Y.J., I.K. Han, J.H. Woo, H.J. Lee, K. Jang, K.H. Myung, and Y.S. Kim. 1997. Compensatory growth in dairy heifers: The effect of a compensatory growth pattern on growth rate and lactation performance. J. Dairy Sci. 80: 519–524.

Clare, D.A. and H.E. Swaisgood. 2000. Bioactive milk peptides: A prospectus. J. Dairy Sci. 83: 1187–1195.

Clark, J.C. 1998. The mammary gland as bioreactor: Expression, processing, and production of recombinant proteins. J. Mammary Gland Biol. Neoplasia 3: 337–350.

Clemmons, D.R. 1998. Role of insulin-like growth factor binding proteins in controlling IGF actions. Mol. Cell. Endocrinol. 140: 19–24.

Cohick, W.S. 1998. Role of the insulin-like growth factors and their binding proteins in lactation. J. Dairy Sci. 81: 1769–1777.

Collier, R. 1985. Nutritional, Metabolic, and Environmental Aspects of Lactation. In Lactation, edited by B.L. Larson. Ames: Iowa State University Press.

Collier, R.J., M.A. Miller, J.R. Hildebrandt, A.R. Torkelson, T.C. White, K.S. Madsen, J.L. Vicini, P.J. Eppard, and G.M. Lanza. 1991. Factors affecting insulin-like growth factor-I concentrations in bovine milk. J. Dairy Sci. 74: 2905–2911.

Collier, R.J., J.C. Byatt, S.C. Denham, P.J. Eppard, A.C. Fabellar, R.L. Hintz, M.F. McGrath, C.L. McLaughlin, J.K. Shearer, J.J. Veenhuizen, and J.L. Vicini. 2001. Effects of sustained release bovine somatotropin (Sometribove) on animal health in commercial dairy herds. J. Dairy Sci. 84: 1098–1108.

Convey, E.M. 1974. Serum hormone concentrations in ruminants during mammary growth, lactogenesis, and lactation: A review. J. Dairy Sci. 57: 905–914.

Cowie, A.T. 1969. Lactogenesis. In The Initiation of Milk Secretion at Parturition, edited by M. Reynolds and S. J. Folley, pp. 157–169. Philadelphia: University of Pennsylvania Press.

Cowie, A.T. and J.S. Tindal. 1971. The Physiology of Lactation. Monographs of the Physiological Society Number 22, edited by H. Davson, A.D.M. Greenfield, R. Whittam, and G.S. Brindley. Baltimore: Williams and Wilkins Company. 392 pages.

Cowie, A.T., J.S. Tindal, and A. Yokoyama. 1966. The induction of mammary growth in the hypophysectomized goat. J. Endocrinol. 34: 185–195.

Cowie, A.T., I.A. Forsyth, and I.C. Hart. 1980. Hormonal Control of Lactation. Heidelberg: Springer–Verlag Berlin.

D

Dahl, G.E., T.H. Elsasser, A.V. Capuco, R.A. Erdman, and R.R. Peters. 1997. Effects of long day photoperiod on milk yield and circulating insulin-like growth factor-1 J. Dairy Sci. 80: 2784–2789.

Dahl, G.E., B.A. Buchanan, and H.A. Tucker. 2000. Photoperiodic effects on dairy cattle: A review. J. Dairy Sci. 83: 885–893.

Daniel, C.W., S. Robinson, and G.B. Silberstein. 1996. The role of TGF-β in patterning growth of the mammary ductal tree. J. Mammary Gland Biol. Neoplasia 1: 331–341.

Das, R. and B. K. Vonderhaar. 1997. Prolactin as a mitogen in mammary cells. J. Mammary Gland Biol. Neoplasia 2: 29–39.

Davis, C.L. and D.E. Bauman. 1974. General metabolism associated with the synthesis of milk. In Lactation: A Comprehensive Treatise, vol. 2, edited by B.L. Larson and V.R. Smith. New York and London: Academic Press.

Davis, C.L. and J. Drackley, 1998. The Development, Nutrition, and Management of the Young Calf. Ames: Iowa State University Press.

Daxenberger, A., B.H. Brier, and H. Sauerwein. 1998. Increased milk levels of insulin-like growth factor 1 (IGF-I) for identification of bovine somatotropin (bST) treated cows. Analyst 123: 2425–2435.

Dunbar, M.E. and J.L. Wysolmerski. 1999. Parathyroid hormone-related protein: A developmental regulatory molecule necessary for mammary gland development. J. Mammary Gland Biol. Neoplasia 4: 21–34.

Dunbar, M.E., P. Young, J. Zhang, J. McCaughern-Carucci, B. Lanske, J.J. Orloff, A. Karaplis, G. Cunha, and J.J. Wysolmerski. 1998. Stromal cells are critical targets in the regulation of mammary ductal morphogenesis by parathyroid hormone related protein. Dev. Biol. 203: 75–89.

Dyer, C.J., J.M. Simmons, R.L. Matteri, and D.H. Keisler. 1997. Leptin receptor mRNA is expressed in ewe anterior pituitary and adipose tissues and is differentially expressed in hypothalamic regions of well-fed and feed-restricted ewes. Domest. Anim. Endocrinol. 14: 119–128.

E

Ehrhardt, R.A., R.M. Slepetis, J. Seigal-Willott, M.E. Van Amburgh, A.W. Bell, and Y.R. Boisclair. 2000. Development of a specific radioimmunoassay to measure physiological changes of circulating leptin in cattle and sheep. J. Endocrinol. 166: 519–528.

Ellis, L.A., A.M. Mastro, and M.F. Picciano. 1996. Milk-borne prolactin and neonatal development. J. Mammary Gland Biol. 1: 259–269.

Ellis, S. 1998. Mechanisms controlling ductal morphogenesis in the ruminant mammary gland. Ph.D. Dissertation, Virginia Polytechnic Institute and State University, Blacksburg, Virginia.

Ellis, S., T.B. McFadden, and R.M. Akers. 1998. Prepubertal ovine mammary development is unaffected by ovariectomy. Domest. Anim. Endocrinol. 15: 217–225.

Ellis, S., S. Purup, K. Sejrsen, and R.M. Akers. 2001. Growth and morphogenesis of epithelial cell organoids from peripheral and medial mammary parenchyma of prepubertal heifers. J. Dairy Sci. 83: 952–961.

Enjalbert, F., M.C. Nicot, C. Bayourthe, R. Mancoulon. 2000. Ketone bodies in milk and blood of dairy cows: Relationship between concentrations and utilization for detection of subclinical ketosis. J. Dairy Sci. 84: 583–589.

Etherton, T.D. and D.E. Bauman. 1998. The biology of somatotropin in growth and lactation of domestic animals. Physiol. Rev. 78: 745–761.

F

Filep, R. and R. M. Akers. 2000. Casein secretion and cytological differentiation in mammary tissue from bulls of high or low genetic merit. J. Dairy Sci. 83: 2261–2268.

Flint, D.J. and M. Gardner. 1994. Evidence that growth hormone stimulates milk synthesis by direct action on the mammary gland and that prolactin exerts effects on milk secretion by maintenance of mammary deoxyribonucleic acid content and tight junction status. Endocrinology 135: 1119–1124.

Flint, D.J., E. Tonner, and G.J. Allan. 2000. Insulin-like growth factor binding proteins: IGF-dependent and -independent effects in the mammary gland. J. Mammary Gland Biol. Neoplasia 5: 65–73.

Forsyth, I.A. 1997. Prolactin, growth hormone and placental lactogen: A historical perspective. J. Mammary Gland Biol. Neoplasia 2: 3–6.

Fowler, P.A., C.H. Knight, G.G. Cameron, and M.A. Foster. 1990a. Use of magnetic resonance imaging in the study of goat mammary glands in vivo. J. Reprod. Fert. 89: 359–366.

Fowler, P.A., C.H. Knight, G.G. Cameron, and M.A. Foster. 1990b. In vivo studies of mammary development in the goat using magnetic resonance imaging (MRI). J. Reprod. Fert. 89: 367–375.

Fowler, P.A., C.H. Knight, and M.A. Foster. 1991. In vivo magnetic resonance imaging studies of mammogenesis in non-pregnant goats treated with exogenous steroids. J. Dairy Res. 58: 151–157.

Friend, T.H., F.C. Gwazdauskas, and C.E. Polan. 1979. Change in adrenal response from free stall competition. J. Dairy Sci. 62: 885–893.

G

Gay, M. 1994. Colostrum research says feed 4 quarts for healthier calves. Hoard's Dairyman 139: 256.

Geishauser, T., K. Leslie, J. Tenhag, and A. Bashiri. 2000. Evaluation of eight cow-side ketone tests in milk for detection of subclinical ketosis in dairy cows. J. Dairy Sci. 83: 296–299.

Glimm, D.R., V.E. Baracos, and J.J. Kennelly. 1992. Northern and in situ hybridization analyses of the effect of somatotropin on bovine mammary gene expression. J. Dairy Sci. 75: 2687–2708.

Goffin, V. and P. Kelly. 1997. The prolactin/growth hormone receptor family: Structure/Function relationships. J. Mammary Gland Biol. Neoplasia 2: 7–18.

Goodman, G.T., R.M. Akers, K.H. Friderici, and H.A. Tucker. 1983. Hormonal regulation of α-lactalbumin secretion from bovine mammary tissue cultured in vitro. Endocrinology 112: 1324–1330.

Groenewegen, P.P., B.W. McBride, and T.H. Elsasser. 1990. Bioactivity of milk of bST-treated cows. J. Nutrition 120: 414–520.

Grossman, M., S.M. Hartz, and W.J. Koops. 1999. Persistency of lactation yield: A novel approach. J. Dairy Sci. 82: 2192–2197.

Grosvenor, C.E., M.F. Picciano, and C.R. Baumrucker. 1993. Hormones and growth factors in milk. Endocr. Rev. 14: 710–727.

Guidry, A.J. 1985. Mastitis and the immune system of the mammary gland. In Lactation, edited by B.L. Larson. Ames: Iowa State University Press.

Gustafsson, J.Å 1999. Estrogen receptor β—A new dimension in estrogen mechanism of action. J. Endocrinol. 163: 379–383.

Gwazdauskas, F.C. 2002. Effects on health and milk production. In Encyclopedia of Dairy Science. Academic Press (in press).

H

Hadsell, D.L. and S.G. Bonnette. 2000. IGF and insulin action in the mammary gland: Lessons from transgenic and knockout models. J. Mammary Gland Biol. Neoplasia 5: 19–30.

Hansen, R.K. and M.J. Bissell. 2000. Tissue architecture and breast cancer: The role of extracellular matrix and steroid hormones. Endocrine-Related Cancer 7: 95–113.

Hanwell, A. and M. Peaker. 1977. Physiological effects of lactation on the mother. Symp. Zool. Soc. Lond. 41: 297–312.

Hauser, S.D., M.F. McGrath, R.J. Collier, and G.G. Krivi. 1990. Cloning and in vivo expression of bovine growth hormone receptor mRNA. Mol. Cell. Endocrinol. 72: 187–200.

Heald, C.W. 1985. Milk collection. In Lactation, edited by B.L. Larson. Ames: Iowa State University Press.

Heald, C.W. and R.G. Saacke. 1972. Cytological comparison of milk protein synthesis of rat mammary tissue in vivo and in vitro. J. Dairy Sci. 55: 621–628.

Heinrichs, A.J. 1996. The importance of heifer raising to a profitable dairy farm. NRAES 74: 1–6

Henninghausen, L., G.W. Robinson, K. Wagner, and X. Liu. 1997. Developing a mammary gland is a Stāt affair. J. Mammary Gland Biol. Neoplasia 2: 365–371.

Hervey, G.R. 1959. The effect of lesions in the hypothalamus in parabiotic rats. J. Physiol. 145: 336–352.

Hoffman, P.C. 1997. Optimum body size of Holstein replacement heifers. J. Anim. Sci. 75: 836–845

Hohenboken, W.D., J. Foldager, J. Jensen, P. Madsens, and B.B. Andersen. 1995. Breed and nutritional effects and interactions on energy intake, production, and efficiency of nutrient utilization in young bulls, heifers, and lactating cows. Acta Agric. Scand. Sect. A. Anim. Sci. 45:92–103.

Houseknecht, K.L., C.A. Baile, R.L. Matteri, and M.E. Spurlock. 1998. The biology of leptin: A review. J. Anim. Sci. 76: 1405–1420.

Hovey, R.C., H.W. Davey, D.D.S. Mackenzie, and T.B. McFadden. 1998. Ontogeny and epithelial-stromal interactions regulate IGF expression in the ovine mammary gland. Mol. Cell. Endocrinol. 136: 139–144.

Hovey, R.C., T.B. McFadden, and R.M. Akers. 1999. Regulation of mammary gland growth and morphogenesis by the mammary fat pad: A species comparison. J. Mammary Gland Biol. Neoplasia 4: 53–68.

Hovey, R.C., H.W. Davey, B.K. Vonderhaar, D.D.S. Mackenzie, and T.B. McFadden. 2001. Paracrine action of keratinocyte growth factor (KGF) during ruminant mammogenesis. Mol. Cell. Endocrinol. 181: 47–56.

Hynes, N.E., N. Cella, and M. Wartmann. 1997. Prolactin mediated intracellular signaling in mammary epithelial cells. J. Mammary Gland Biol. Neoplasia 2: 19–27.

I

Ingvartsen, K.L. and J.B. Andersen. 2000. Integration of metabolism and intake regulation: A review focusing on periparturient animals. J. Dairy Sci. 83: 1573–1597.

Ingvartsen, K.L. and K. Sejrsen. 1995. Effect of immunization against somatostatin (SS) in cattle: A review of performance, carcass composition and possible mode of action. Acta. Agric. Scand., Sect. A., Anim. Sci. 45: 124–131.

J

James, R.E. 2001. Growth standards and nutrient requirements for dairy heifers—weaning to calving. Adv. Dairy Technol. 13: 63–77.

Jenness, R. 1974. The composition of milk. In Lactation: A Comprehensive Treatise, vol. III, edited by B.L. Larson and V.R. Smith, pp. 3–96. New York and London: Academic Press.

Jenness, R. 1985. Biochemical and nutritional aspects of milk and colostrums. In Lactation, edited by B.L. Larson. Ames: Iowa State University Press.

Jobst, P.M., S.D. Berry, M.L. McGilliard, D. Ayares, D.A. Henderson, W.E. Beal, and R.M. Akers. 2001. Local expression of IGF-I and IGFBP-3 mRNA in mammary tissue of prepubertal heifers after treatment with growth hormone. J. Dairy Sci. 84(Suppl. 1): 314.

Johnson, I.D. 1988. The effect of pubertal nutrition on lactation performance by dairy cows. In Nutrition and Lactation in the Dairy Cow, edited by P.C. Garnsworthy. London: Butterworth.

Jones, G.M., R.E. Pearson, G.A. Clabaugh, and C.W. Heald. 1984. Relationships between somatic cell counts and milk production. J. Dairy Sci. 67: 1823–1831.

K

Kamalati, T., B. Niranjan, J. Yant, and L. Buluwela. 1999. HGF/SF in mammary growth and morphogenesis: In vitro and in vivo models. J. Mammary Gland Biol. Neoplasia 4: 69–77.

Karatzas, C.N. and J.D. Turner. 1997. Toward altering milk composition by genetic manipulation: Current status and challenges. J. Dairy Sci. 80: 2225–2232.

Kazmer, G.W., M.A. Barnes, R.M. Akers, and R.E. Pearson. 1986. Effect of genetic selection for milk yield and increased milking frequency on plasma growth hormone and prolactin concentrations in Holstein cows. J. Anim. Sci. 63: 1220–1227.

Keely, P.J., J.E. Wu, and S.A. Santoro. 1995. The spatial and temporal expression of the $\alpha_2\beta_1$ integrin and its ligands, collagen I, collagen IV, and laminin, suggest important roles in mouse mammary morphogenesis. Differentiation 59: 1–13.

Kelly, M.L., E.S. Kolver, D.E. Bauman, M.E. Van Amburgh, and L.D. Muller. 1998. Effect of intake of pasture on concentrations of conjugated linoleic acid in milk lactating cows. J. Dairy Sci. 81: 1630–1636.

Kennedy, G.C. 1953. The role of depot fat in the hypothalamic control of feed intake in the rat. Proc. R. Soc., Ser. B 139: 578–592.

Keown, J.F. and R.W. Everett. 1986. Effect of days carried calf, days dry and weight of 1st calf heifers on yield. J. Dairy Sci. 69: 1891–1901.

Kertz, A.F., L.R. Prewitt, and J.M. Ballan. 1987. Increased weight gain and effect on growth parameters of Holstein heifer calves from 3 to 12 months of age. J. Dairy Sci. 70: 1612–1622.

Keys, J.E., A.V. Capuco, R.M. Akers, and J. Djiane. 1989. Comparative study of mammary gland development and differentiation between beef and dairy heifers. Domest. Anim. Endocrinol. 6: 311–319.

Keys, J.E., J.P. Van Zyl, and H.M. Farrell, Jr. 1997. Effect of somatotropin and insulin-like growth factor-I on milk lipid and protein synthesis in vitro. J. Dairy Sci. 80: 37–45.

Kingwill, R.G., F.H. Dodd, and F.K. Neave. 1979. Machine milking and mastitis. In Machine Milking, edited by C.C. Thiel and F.H. Dodd, Technical Bulletin 1, National Institute for Research in Dairying, Reading, England, Hannah Research Institute, Ayr, Scotland.

Kleinberg, D.L., M. Feldman, and W. Ruan. 2000. IGF-I an essential factor in terminal end bud formation and ductal morphogenesis. J. Mammary Gland Biol. 5: 7–19.

Knight, C.H. and M. Peaker. 1982. Development of the mammary gland. J. Reprod. Fert. 65: 521–536.

Koff, M.D. and K. Plaut. 1995. Detection of transforming growth factor-alpha like messenger RNA in the bovine mammary gland. J. Dairy Sci. 78: 1903–1908.

Koldovskkÿ, O. 1996. The potential physiological significance of milk-borne hormonally active substances for the neonate. J. Mammary Gland Biol. 1: 317–323.

Koprowski, J.A. and H.A. Tucker. 1973. Bovine serum growth hormone, corticoids and insulin during lactation. Endocrinology 93: 645–651.

Kronfeld, D.S. 1982. Major metabolic determinants of milk volume, mammary efficiency, and spontaneous ketosis in dairy cows. J. Dairy Sci. 65: 2204–2212.

L

Lacasse, P., V.C. Farr, S.R. Davis, and C.G. Prosser. 1996. Local secretion of nitric oxide and the control of mammary blood flow. J. Dairy Sci. 79: 1369–1374.

Lamb, G.C., B.L. Miller, J.M. Lynch, D.M. Grieger, J.S. Stevenson, and M.C. Lucy. 1999. Suckling reinitiated milk secretion in beef cows after an early postpartum hiatus of milking or suckling. J. Dairy Sci. 82: 1489–1496.

Lammers, B.P, A.J. Heinrichs, and R.S. Kensinger. 1999a. The effects of accelerated growth rate and estrogen implants in prepubertal Holstein heifers on growth, feed efficiency and blood parameters. J. Dairy Sci. 82: 1746–1752.

Lammers, B.P, A.J. Heinrichs, and R.S. Kensinger. 1999b. The effects of accelerated growth rate and estrogen implants in prepubertal Holstein heifers on estimates of mammary development and subsequent reproduction and milk production. J. Dairy Sci. 82: 1753–1764.

Larson, B.L. (ed.). 1985. Lactation. Ames: Iowa State University Press.

Larson, B.L. 1985. Biosynthesis and cellular secretion of milk. In Lactation, edited by B.L. Larson. Ames: Iowa State University Press.

Laud, K., I. Gourdou, L. Belair, D.H. Keisler, and J. Djiane. 1999. Detection and regulation of leptin receptor mRNA in ovine mammary epithelial cells during pregnancy and lactation. FEBS Lett. 463: 194–198.

Lefcourt, A.M. and R.M. Akers. 1983. Is oxytocin really necessary for efficient milk removal in dairy cows? J. Dairy Sci. 66: 2251–2259.

Lefcourt, A.M., J. Bitman, D.L. Wood, and R.M. Akers. 1995. Circadian and ultradian rhythms of peripheral growth hormone concentrations in lactating dairy cows. Domest. Anim. Endocrinol. 12: 247–256.

Le Roith, D., C. Bondy, S. Yakar, J. Lui, and A. Butler. 2001. The somatomedin hypothesis 2001. Endocr. Rev. 22: 53–74.

Littledike, E.T., J.W. Young, and D.C. Beitz. 1981. Common metabolic diseases of cattle: Ketosis, milk fever, grass tetany, and downer cow complex. J. Dairy Sci. 64: 1465–1482.

Lyons, W.R., C.H. Li, and R.E. Johnson. 1958. The hormonal control of mammary growth. Recent Prog. Horm. Res. 14: 219–254.

M

Malven, P.V. 1977. Prolactin and other protein hormones in milk. J. Anim. Sci. 45: 609–616.

Mantysaari, P., K.L. Ingvartsen, V. Toivonen, and K. Sejrsen. 1995. The effects of feeding level and nitrogen source of the diet on mammary development and plasma hormone concentrations of prepubertal heifers. Acta. Agric. Scand. Sect. A Anim. 45: 236–242.

Mather, I.H. 2000. A review and proposed nomenclature for major proteins of the milk-fat globule membrane. J. Dairy Sci. 83: 203–247.

Mather, I.H. and T.W. Keenan. 1998. Origin and secretion of milk lipids. J. Mammary Gland Biol. Neoplasia 3: 259–273.

McFadden, T.B., R.M. Akers, and G.W. Kazmer. 1987. Alpha-lactalbumin in bovine serum: Relationships with udder development and function. J. Dairy Sci. 70: 259–264.

McFadden, T.B., R.M. Akers, and A.V. Capuco. 1988. Relationships of milk proteins in blood with somatic cell counts in milk of dairy cows. J. Dairy Sci. 71: 826–834.

McFadden, T.B., T.E. Daniel, and R.M. Akers. 1990. Effects of plane of nutrition, growth hormone, and unsaturated fat on mammary growth in prepubertal lambs. J. Anim. Sci. 68: 3171–3179.

Meisel, H. 1997. Biochemical properties of regulatory peptides derived from milk proteins. Biopolymers 43: 119–128.

Mepham, T.B. 1983. Physiological aspects of lactation. In Biochemistry of Lactation, edited by T.B. Mepham, pp. 4–28. B.V. Amsterdam: Elsevier Science Publishers, 1983.

Miettinen, M., V. Isomaa, H. Peltoketo, D. Ghosh, and P. Vihko. 2000. Estrogen metabolism as a regulator of estrogen action in the mammary gland. J. Mammary Gland Biol. Neoplasia 5: 259–270.

N

Nandi, S. 1958. Endocrine control of mammary gland development and function in the C3H/He Crgl mouse. J. Natl. Cancer Inst. 21: 1039–1360.

National Mastitis Council. 1996. Current Concepts of Bovine Mastitis, 4th edition. Arlington, Virginia: National Mastitis Council.

Neijenhuis, F., H.W. Barkema, H. Hogeveen, and J.P.T.M. Noordhuizen. 2000. Classification and longitudinal examination of callused teat ends in dairy cows. J. Dairy Sci. 83: 2795–2804.

Nickerson, S.C. and R.M. Akers. 1983. Effect of prepartum blockade of microtubule formation on ultra-structural differentiation of the mammary epithelium in Holstein heifers. Int. J. Biochem. 15: 777–788.

Nickerson, S.C. and R.M. Akers. 1984. Biochemical and ultrastructural aspects of milk synthesis and secretion. Int. J. Biochem. 16: 855–865.

Nickerson, S.C. and C.W. Heald. 1981. Histopathologic response of the bovine mammary gland to experimentally induced *Staphylococcus aureus* infection. Am. J. Vet. Res. 42: 1351–1355.

Nickerson, S.C. and J.W. Pankey. 1983. Cytological observations of the bovine teat end. Am. J. Vet. Res. 44: 1433–1441.

Nicoll, C.S. 1997. Cleavage of prolactin by its target organs and the possible significance of this process. J. Mammary Gland Biol. Neoplasia 2: 81–89.

Nocek, J.E. 1997. Bovine acidosis: Implications on laminitis. J. Dairy Sci. 80: 1005–1028.

Nostrand, S.D., D.M. Galton, H.N. Erb, and D.E. Bauman. 1991. Effects of daily exogenous oxytocin on lactation milk yield and composition. J. Dairy Sci. 74: 2119–2127.

O

Oftendal, O.T. 1984. Milk composition, milk yield and energy output at peak lactation: A comparative review. Symp. Zool. Soc. Lond. 51: 33–85.

Oftendal, O.T. 1997. Lactation in whales and dolphins: Evidence of divergence between baleen- and toothed-species. J. Mammary Gland Biol. Neoplasia 2: 205–230.

P

Pajor, E.A., J. Rushen, and A.M. de Passille. 2000. Cow comfort, fear, and productivity. In Dairy Housing and Equipment Systems. Ithaca, New York: Natural Resource, Agriculture and Engineering Service.

Parameswaran, S.V., A.B. Steffens, G.R. Hervey, and L. DeRuiter. 1977. Involvement of a humoral factor in the regulation of body weight in parabiotic rats. Am. J. Physiol. 232: R150–R157.

Park, C.S., G.M. Erickson, Y.J. Choi, and G.O. Marx. 1987. Effects of compensatory growth on regulation of growth and lactation: Response of dairy heifers to a stair-step growth pattern. J. Anim. Sci. 64: 1751–1759.

Park, C.S., R.B. Danielson, B.S. Kreft, S.H. Kim, Y.S. Moon, and W.L. Keller. 1998. Nutritionally directed compensatory growth and effects on lactation potential of developing heifers. 81: 243–249.

Parodi, P.W. 1999. Conjugated linoleic acid and other anticarcinogenic agents of bovine milk fat. J. Dairy Sci. 82: 1339–1349.

Parrish, D.B., G.H. Wise, J.S. Hughes, and F.W. Atkeson. 1948. Properties of colostrum of the dairy cow. II. Effect of prepartal rations upon the nitrogenous constituents. J. Dairy Sci. 33: 457–468.

Parrish, D.B., G.H. Wise, J.S. Hughes, and F.W. Atkeson. 1950. Properties of colostrum of the dairy cow. V. Yield, specific gravity and concentrations of total solids and its various components of colostrum and early milk. J. Dairy Sci. 33: 457–466.

Peaker, M. and C.J. Wilde. 1996. Feedback control of milk secretion from milk. J. Mammary Gland Biol. 1: 307–315.

Peters, R.R., L.T. Chapin, K.B. Leining, and H.A. Tucker. 1978. Supplemental lighting stimulates growth and lactation in cattle. Science 199: 911–912.

Petitclerc, D., L.T. Chapin, and H.A. Tucker. 1984. Carcass composition and mammary development responses to photoperiod and plane of nutrition in Holstein heifers. J. Anim. Sci. 58: 892–898.

Plath-Gabler, A., C. Gabler, F. Sinowatz, B. Berisha, and D. Schams. 2001. The expression of the IGF family and GH receptor in the bovine mammary gland. J. Endocrinol. 168: 39–48.

Plaut, K. 1993. Role of epidermal growth factor and transforming growth factors in mammary development and lactation. J. Dairy Sci. 76: 1526–1538.

Plaut, K. and R.L. Maple. 1995. Characterization of binding of transforming growth factor-beta-I to bovine mammary membranes. J. Dairy Sci. 78: 1463–1469.

Plaut, K., D.E. Bauman, N. Agergaard, and R.M. Akers. 1987. Effect of exogenous prolactin administration on lactational performance of dairy cows. Domest. Anim. Endocrinol. 4: 279–290.

Politis, I. 1996. Plasminogen activator system: Implications for mammary cell growth and involution. J. Dairy Sci. 79: 1097–1107.

Pope, G.S. and J.K. Swinburne. 1980. Reviews in the progress of dairy science. Hormones in milk: Their physiological significance and value as diagnostic aids. J. Dairy Res. 47: 427–449.

Powers, C.J., S.W. McLeskey, and A. Wellstein. 2000. Fibroblast growth factors, their receptors and signaling. Endocrine-Related Cancer 7: 165–197.

Prosser, C.G. 1996. Insulin-like growth factors in milk and mammary gland. J. Mammary Gland Biol. Neoplasia 3: 297–306.

Prosser, C.G., S.R. Davis, V.C. Farr, and P. Lacasse. 1996. Regulation of blood flow in the mammary microvasculature. J. Dairy Sci. 79: 1184–1197.

Purup, S., K. Sejrsen, and R.M. Akers. 1993a. Influence of estradiol on insulin-like growth factor I (IGF-I) stimulation of DNA synthesis in vitro in mammary gland explants from intact and ovariectomized prepubertal Holstein heifers. Livest. Prod. Sci. 35: 182.

Purup, S., K. Sejrsen, J. Foldager, and R.M. Akers. 1993b. Effect of exogenous bovine growth hormone and ovariectomy on prepubertal mammary growth, serum hormones and acute in-vitro proliferative response of mammary explants from Holstein heifers. J. Endocrinol. 139: 19–26.

Purup, S., K..Sejrsen, and R.M. Akers. 1995. Effect of bovine GH and ovariectomy on mammary sensitivity to IGF-I in prepubertal heifers. J. Endocrinol. 144: 153–158.

Purup, S., M. Vestergaard, and K. Sejrsen. 2000a. Involvement of growth factors in the regulation of pubertal mammary growth in cattle. Adv. Exp. Med. Biol. 480: 27–43.

Purup, S., M. Vestergaard, M.S. Weber, K. Plaut, R.M. Akers, and K. Sejrsen. 2000b. Local regulation of pubertal mammary growth in heifers. J. Anim. Sci. 78(Suppl. 3): 36–47.

Q

Quigley, J.D. and J.J. Drewy. 1998. Nutrient and immunity transfer from cow to calf pre- and postcalving. J. Dairy Sci. 81: 2779–2790.

R

Radcliff, R.P., M.J. VandeHaar, A.L. Skidmore, L.T. Chapin, B.R. Radke, J.W. Lloyd, E.P. Stanisiewski, and H.A. Tucker. 1997. Effects of diet and bovine somatotropin on heifer growth and development. J. Dairy Sci. 80: 1996–2003.

Radcliff, R.P., M.J. VandeHaar, L.T. Chapin, T.E. Pilbeam, D.K. Beede, E.P. Stanisiewski, and H.A. Tucker. 2000. Effects of prepubertal feeding level and injection of bovine somatotropin on first lactation milk yields of Holstein cows. J. Dairy Sci. 83: 23–29.

Riddle, O., R.W. Bates, and S.W. Dykshorn. 1933. The preparation, identification and assay of prolactin—a hormone of the anterior pituitary. Am. J. Physiol. 105: 191–216.

Rosen, J.M., S.L. Wyszomierski, and D. Hadsell. 1999. Regulation of milk protein gene expression. Annu. Rev. Nutr. 19: 407–436.

S

Sanchez-Barcelo, E.J., M.D. Mediavilla, S.A. Zinn, B.A. Buchanan, L.T. Chapin, and H.A. Tucker. 1991. Melatonin suppression of mammary growth in heifers. Biol. Reprod. 33: 324–334.

Schanbacher, F.L., R.S. Talhouk, and F.A. Murray. 1997. Biology and origin of bioactive peptides in milk. Livest. Prod. Sci. 50: 105–123.

Sejrsen, K. and J. Foldager. 1992. Mammary growth and milk production capacity of replacement heifers in relation to diet energy concentration and plasma hormone levels. Anim. Sci. 42: 99–107.

Sejrsen, K. and S. Purup. 1997. Influence of prepubertal feeding level on milk yield potential of dairy heifers: A review. J. Anim. Sci. 75: 828–835.

Sejrsen, K., J.T. Huber, H.A. Tucker, and R.M. Akers. 1982. Influence of plane of nutrition on mammary development in pre- and post-pubertal heifers. J. Dairy Sci. 65: 793–800.

Sejrsen, K., J. Foldager, M.T. Sorensen, R.M. Akers, and D.E. Bauman. 1986. Effect of exogenous bovine somatotropin on pubertal mammary development in heifers. J. Dairy Sci. 69: 1528–1535.

Sejrsen, K., S. Purup, M. Vestergaard, M.S. Weber, and C.H. Knight. 1999. Growth hormone and mammary development. Domest. Anim. Endocrinol. 17: 117–129.

Sejrsen, K., S. Purup, M. Vestergard, and J. Foldager. 2000. High body weight gain and reduced bovine mammary growth: Physiological basis and implications for milk yield potential. Domest. Anim. Endocrinol. 19: 93–104.

Sheffield, L.G. 1988. Organization and growth of mammary epithelia in the mammary fat pad. J. Dairy Sci. 71: 2855–2874.

Sheffield, L.G. 1998. Hormonal regulation of epidermal growth factor receptor content and signaling in bovine mammary tissue. Endocrinology 139: 4568–4575.

Shennan, D.B. and M. Peaker. 2000. Transport of milk constituents by the mammary gland. Physiol. Rev. 80: 925–951.

Sinha, Y. 1995. Structural variants of prolactin: Occurrence and physiological significance. Endocr. Rev. 16: 354–369.

Sinha, Y.N. and H.A. Tucker. 1969. Mammary development and pituitary prolactin level of heifers from birth through puberty and during estrus cycle. J. Dairy Sci. 52: 507–512.

Sinowatz, F., D. Schams, S. Kolle, A. Plath, D. Lincoln, and M.J. Waters. 2000. Cellular localization of GH receptor in the bovine mammary gland during mammogenesis, lactation and involution. J. Endocrinol. 166: 503–510.

Sirbasku, D.A. 1978. Estrogen induction of growth factors specific for hormone-responsive mammary, pituitary, and kidney tumor cell. Proc. Natl. Acad. Sci. 75: 3786–3790.

Smith, G.H. 1996. TGF-β and functional differentiation. J. Mammary Gland Biol. Neoplasia 1: 343–352.

Smith, J.J., A.V. Capuco, and R.M. Akers. 1987. Quantification of progesterone binding in mammary tissue of pregnant ewes. J. Dairy Sci. 70: 1178–1185.

Smith, J.J., A.V. Capuco, W.E. Beal, and R.M. Akers. 1989. Association of prolactin and insulin receptors with mammogenesis and lobulo-alveolar formation in pregnant ewes. Int. J. Biochem. 21: 73–81.

Smith, J.J., A.V. Capuco, I.H. Mather, and B.K. Vonderhaar. 1993. Ruminants express a prolactin receptor of M(r) 33,000–36,000 in the mammary gland throughout pregnancy and lactation. J. Endocrinol. 139: 37–49.

Smith, K.L. and F.L. Schanbacher. 1973. Hormone induced lactation in the bovine. I. Lactational performance following injections of 17β-estradiol and progesterone. J. Dairy Sci. 56: 738–745.

Sordillo, L.M., K. Shafer-Weaver, and D. DeRosa. 1997. Immunobiology of the mammary gland. J. Dairy Sci. 80: 1851–1865.

Stelwagen, K. 2001. Effect of milking frequency on mammary functioning and shape of the lactation curve. J. Dairy Sci. 84(E. Suppl.): E204–E211.

Stricker, P. and R. Grueter. 1928. Action du lobe anterieur de l'hypophyse sur la montée laiteuse. CR Soc. Biol. (Paris) 99: 1978–1980.

Strong, C.R. and R. Dils. 1972. Fatty acid biosynthesis in rabbit mammary gland during pregnancy and early lactation. Biochem. J. 128: 1303–1309.

Swett, W.W., P.C. Underwood, C.A. Matthews, and R.R. Graves. 1942. Arrangement of the tissues by which the cow's udder is suspended. J. Agric. Res. 65: 19–42.

Sympson, C.J., R.S. Talhouk, C.M. Alexander, J.R. Chin, S.M. Clift, M.J. Bissell, and Z. Werb. 1994. Targeted expression of stromelysin-1 in mammary gland provides evidence of a role for an intact basement membrane for tissue specific gene expression. J. Cell Biol. 125: 681–693.

T

Topper, Y.J. and C.S. Freeman. 1980. Multiple hormone interactions in the developmental biology of the mammary gland. Physiol. Rev. 60: 1049–1106.

Tucker, H.A. 1981. Physiological control of mammary growth, lactogenesis, and lactation. J. Dairy Sci. 64:1403–1421.

Tucker, H.A. 1985. Endocrine and neural control of the mammary gland. In Lactation, edited by B.L. Larson. Ames: Iowa State University Press.

Tucker, H.A 1987. Quantitative estimates of mammary growth during various physiological states: A review. J. Dairy Sci. 70: 1958–1966.

Tucker, H.A. 1994. Lactation and its hormonal control. In The Physiology of Reproduction, 2nd edition, edited by E. Knobil and J. D. Neill, pp. 1065–1098. New York: Raven Press.

Tucker, H.A. 2000. Hormones, mammary growth, and lactation: A 41-year perspective. J. Dairy Sci. 83: 874–884.

Twining, S.S. 1994. Regulation of proteolytic activity in tissues. Crit. Rev. Biochem. Mol. Biol. 29: 315–343.

V

Van Amburgh, M. and J. Tikofsky. 2001. The advantages of "accelerated growth" in heifer rearing. Adv. Dairy Tech. 13: 79–97.

Van Amburgh, M.E., D.M. Galton, D.E. Bauman, R.W. Everett, D.G. Fox, L.E. Chase, and H.N. Erb. 1998. Effects of three prepubertal body growth rates on performance of Holstein heifers during first lactation. J. Dairy Sci. 81: 527–538.

W

Waldo, D.R., A.V. Capuco, and C.E. Rexroad, Jr. 1998. Milk production of Holstein heifers fed either alfalfa or corn silage diets at two rates of daily gain. J. Dairy Sci. 81: 756–764.

Weber, M.S., S. Purup, S. Vestergaard, S.E. Ellis, J. Sondergaard, R.M. Akers, and K. Sejrsen. 1999. Contribution of insulin-like growth factor-I (IGF-I) and IGF-binding protein-3 (IGFBP-3) to mitogenic activity in bovine mammary extracts and serum. J. Endocrinol. 161: 365–373.

Weber, M.S., S. Purup, M. Vestergaard, R.M. Akers, and K. Sejrsen. 2000a. Regulation of local synthesis of insulin-like growth factor-I and binding proteins in mammary tissue. J. Dairy Sci. 83: 30–37.

Weber, M.S., S. Purup, M. Vestergaard, R.M. Akers, and K. Sejrsen. 2000b. Nutritional and soma-totropin regulation of the mitogenic response of mammary cells to mammary tissue extracts. Domest. Anim. Endocrinol. 18: 159–164.

Wilde, C.J., C.H. Knight, and D.J. Flint. 1999. Control of milk secretion and apoptosis during mammary involution. J. Mammary Gland Biol. 4: 129–136.

Woodward, T.L., W.E. Beal, and R.M. Akers. 1993. Cell interactions in initiation of mammary epithelial proliferation by oestradiol and progesterone in prepubertal heifers. J. Endocrinol. 136: 149–157.

Woodward, T.L, J.W. Xie, and S.Z. Haslam. 1998. The role of mammary stroma in modulating the proliferative response to ovarian hormones in the normal mammary gland. J. Mammary Gland Biol. Neoplasia 3: 117–131.

Woodward, T.L., J. Xie, J.L. Fendrick, and S.Z. Haslam. 2000. Proliferation of mouse mammary epithelial cells in vitro: Interactions among epidermal growth factor, insulin-like growth factor-I, ovarian hormones, and extracellular matrix proteins. Endocrinology 141: 3578–3586.

Y

Yang, J., J.J. Kennelly, and V.E. Barcos. 2000a. The activity of transcription factor Stat5 responds to prolactin, growth hormone, and IGF-I in rat and bovine mammary explant culture. J. Anim. Sci. 78: 3114–3125.

Yang, J., J.J. Kennelly, and V.E. Barcos. 2000b. Physiological levels of Stat5 DNA binding activity and protein in bovine mammary gland. J. Anim. Sci. 78: 3126–3134.

Z

Zhang, J., D.G. Grieve, R.R. Hacker, and J.H. Burton. 1995. Effect of dietary protein percentage and β-agonist administered to prepubertal ewes on mammary gland growth and milk secretion. J. Anim. Sci. 73: 2655–2661.

Zhao, X., B.W. McBride, L.M. Trouten-Radford, L. Golfman, and J.H. Burton. 1994. Somatotropin and insulin-like growth factor-I concentrations in plasma and milk after daily injections of sustained-release exogenous somatotropin administrations. Domest. Anim. Endocrinol. 11: 209–216.

Glossary

A

Adrenergic receptors are cellular proteins that specifically bind norepinephrine and epinephrine and similar molecules. They mediate neural signaling in the CNS and hormonal signaling via epinephrine in some peripheral tissues. So named because epinephrine and norepinephrine were originally called adrenalin and noradrenalin.

Allometry indicates a rate of growth of a tissue or organ that is faster than that of the body as a whole.

α–lactalbumin is a specific milk protein synthesized and secreted by alveolar epithelial cells. The protein is also a part of the lactose synthetase enzyme. It appears in the whey fraction of milk.

Alveoli are the multicellular, spherical, hollow units of the mammary gland responsible for synthesis and secretion of milk. Milk is secreted into and stored in the internal lumenal space of the alveoli between milking and suckling episodes.

B

bST or bovine somatotropin is also called *growth hormone* (GH) and is produced in the anterior pituitary gland. A common terminology is to use a lowercase letter to indicate the species of origin (e.g., pST for porcine somatotropin) and a small *r* to indicate that the protein is derived by recombinant DNA technology. For example, rbST stands for recombinant bovine somatotropin or growth hormone.

C

Carcinoma means cancer of the epithelial cells or the epithelium. It is the most common form of cancer in humans.

Caseins constitute the major group of specific milk proteins. They are empirically defined by their precipitation from milk at pH 4.6 and their capacity to produce micelles. Major subtypes include α, β, and γ caseins.

Chaperone is a protein that allows other proteins to avoid alterations in folding or conformation so that the supported protein maintains its functional shape.

Cytology is the study of the structure, organelles, and function of cells.

D

De novo means occurring within a tissue or cell.

Dystocia is a term that means difficult birth and is often used in dairy management-recording schemes to indicate that an animal required assistance with calving.

267

E

Ectoderm is one of three germinal cell layers (ectoderm, mesoderm, and endoderm) in the developing embryo. The mammary gland is derived from the ectoderm.

End buds are the swollen terminal ends of mammary ducts in the mammary glands of peripubertal rodents that are responsible for rapid growth and elongation of the ductular tree. Although alveolar bud-like structures in the mammary glands of peripubertal ruminants serve as sites for rapid growth of mammary ducts, ruminants apparently do not have morphologically similar end buds.

Epidermal growth factor (EGF) and epidermal growth factor receptor (EGFR or ERBB1) make up a family of related growth factors and receptors that are believed to be involved in regulation of normal mammary growth and mammary cancer in many species. Receptors for the family include EGFR and several ligands, including amphiregulin, TGF-α, and ERBB2 (called HER in humans and Neu in rodents). Several family members (ERBB3, EGFR, and TGF-α) are often upregulated in breast cancer.

G

Galactopoiesis refers to the maintenance or continuation of established lactation. For example, secretion of growth hormone is believed to be essential for a successful lactation in ruminants; thus, GH is considered a galactopoietic hormone.

Gel filtration is a separation technique that depends on differences in rate of passage of proteins applied to a column prepared from polymers with varying pore sizes. Smaller proteins enter narrow pores and are more easily retarded than larger proteins.

G protein is a generic term for a member of a large family of guanine triphosphate (GTP) -binding proteins that are important elements for cell signaling. The binding of a hormone or other signaling ligand in this family activates these cellular proteins, many of which act as kinases.

Golgi or Golgi apparatus is the cellular organelle closely linked with packaging and processing (phosphorylation, glycosylation, etc.) of proteins destined to be secreted from the cell. Stacks of Golgi membranes and associated secretory vesicles are abundant in the apical cytoplasm of fully differentiated mammary cells from lactating animals.

Growth factors are small peptides that act by an autocrine, a paracrine, or a classic endocrine loop to stimulate or inhibit the growth and development of tissues and cells.

H

Heparin is a common extracellular proteoglycan with polysaccharide chains linked to a common core protein. Repeating disaccharide units are composed of glucosamine and glucuronic acid.

Holocrine glands are glands that contain secretory cells that accumulate their products and then are sloughed and disrupted to form the secretion from the glands.

Homeorrhresis indicates a state of adjusted or altered physiology to support a particular body activity or function (e.g., lactation or reproduction).

Homeostasis refers to a state of equilibrium or stability of the internal environment of the body.

Hypothalamus is a primitive area of the brain located below the cerebellum and thalamus but superior to the pituitary gland.

I

Immunoglobulins make up a class of blood proteins that include antibodies of several types (IgG, IgA, IgE, IgD, and IgE).

Insulin-like growth factor I (IGF-I) is a small growth factor (~7.5 kD) that appears in circulation largely in response to growth hormone stimulation of the liver. It is a potent stimulator of cell proliferation and is also locally produced in the stromal tissue of the mammary gland.

Insulin-like growth factor–binding proteins (IGFBPs) make up a family of at least six well-characterized proteins and several more distant relatives that can bind IGF-I. They are believed to modulate the biological effectiveness of IGF-I. IGFBP-2, -3, and -4 are especially evident in serum. IGFBPs are also locally produced in many tissues including the mammary gland.

Insulin receptor substrate (IRS) is an intracellular mediator of IGF-I and insulin action. Binding of the proteins to their surface receptors causes autophosphorylation of the receptor and creation of docking sites for IRS family members. Docking of IRS allows further interaction with a cascade of other signaling molecules, including the p85 subunit of PI3K. Downstream effectors of PI3K include the kinases Akt, $p70^{s6}$, and protein kinase C. PI3K is believed to be especially important in regulation of mitogenesis and via the Akt pathway involved in cell survival. For example, the proapoptotic protein BAD, which causes cell death when associated with Bcl2 or Bcl_{xl}, is phosphorylated in the presence of Akt and effectively inhibited. These signaling events explain major attributes of IGF-I in the mammary gland (i.e., stimulation of cell growth and maintenance of cell number).

In vitro means in glass and refers to studies done outside the body with tissues, cells, or cell components in various laboratory settings (e.g., test tubes or culture dishes).

J

Janus protein tyrosine kinases (JAKS) are kinases that become activated following binding of a number of cytokine family members to their respective receptors. They are involved in signal transduction.

L

Lactogenesis refers to the onset of lactation near the time of parturition. It occurs in two phases, with limited structural and functional differentiation of the alveolar epithelial cells after lobulo-alveolar development during gestation, followed by dramatic differentiation and onset of copious milk secretion within days or hours of parturition.

Lobulo-alveolar is a developmental term indicating a structural grouping of several alveoli, their terminal ducts and related common mammary ducts, and surrounding supporting connective tissue.

M

Mammogenic is a term given to substances that stimulate mammary growth and development. For example, the ovarian hormone estrogen is a classic mammogenic hormone.

Mesenchyme indicates tissues or cells derived from the embryonic mesoderm. In the developing mammary gland, the stromal tissue surrounding the epithelial cells contains primitive cells capable of being induced to differentiate into one of several stromal tissue cell types (i.e., endothelial, fibroblast, or adipocytes).

Mitogen-activated protein kinase (MAPK) is a protein kinase that performs a crucial step in relaying signals from the plasma membrane to the nucleus. It is activated by a large array of proliferation- or differentiation-inducing signals from outside the cell.

Multiparous indicates that an animal has had more than one pregnancy and birth.

N

Nicotinamide adenine dinucleotide (NAD) is a molecule that serves as a coenzyme for oxidative pathways (glycolysis, Krebs cycle, electron transport chain). It serves to transfer electrons in oxidation-reduction reactions and is derived from niacin.

Nuclear magnetic resonance imaging (MRI) stands for resonant absorption of electromagnetic radiation at a specific frequency by atomic nuclei in a magnetic field, due to a change in the orientation of their magnetic dipole moments. It is used as an imaging technique for organs and tissues.

O

Opioid peptides are naturally occurring brain proteins involved in regulation of neural activity as well as control of secretion of several hypothalamic and pituitary hormones.

Oxidative phosphorylation is the process of ATP synthesis during which an inorganic phosphate group becomes attached to ADP. It occurs via the electron transport chain in the mitochondria.

P

Parenchyma is the functional portion of a tissue or organ. For example, in the mammary gland the alveoli and mammary ducts are responsible for the synthesis, storage, and transport of milk from the mammary gland.

Passive immunity refers to immunity that is derived from transfer rather that activation of an animal's own immune system. Examples include antibodies passed to the fetus in utero or across the gut of the newborn via colostrums.

Phagocytosis is a term that refers to the amoeboid-like engulfment of extracellular material by one of the immune cells, most often macrophages or neutrophils.

Phosphatidylinositol-3-kinase (PI3K) is an enzyme involved in the synthesis of the phosphoinositide family of lipid second messengers. Members of this family of intracellular signaling molecules are thought to be critical to suppression of signals that can cause apoptosis or programmed cell death.

Phosphorylation refers to addition of a phosphate group to protein by the action of an enzyme called *kinase*.

Primiparous refers to an animal during its first pregnancy or following birth of the first offspring.

Prolactin (Prl) is a hormone produced in the anterior pituitary gland. It is a critical regulator of mammary development in many species and is especially important in the differentiation of the alveolar epithelial cells at the time of parturition or lactogenesis.

Proteoglycan is a molecule consisting of glycosaminoglycan (GAG) chains attached to a core protein.

Proto-oncogene is a normal gene, usually involved in regulation of cell growth, that can be converted into a cancer-stimulating oncogene by mutation.

Pyrimidine is one of two classes of nitrogenous bases found in DNA and RNA. The other is purine.

R

Radioimmunoassay (RIA) is a sensitive assay method used to measure the concentrations of hormones and other factors in biological fluids. The technique depends on the ability to produce antibodies against the substance under study and to label it with a radioisotope.

Ras protein is an example of a large family of GTP-binding proteins that serve to relay signals from cell surface receptors to the cell nucleus. It is named for the ras gene, first identified in viruses that cause sarcoma in rats.

Rough endoplasmic reticulum (RER) is the cellular organelle involved in translation of mRNA for synthesis of proteins for secretion from the cells. Appears as parallel arrays of intracellular membranes studded with ribosomes.

S

SCR gene is the name of the first retroviral oncogene (v-scr) and its precursor proto-oncogene (c-src). The product of the gene is a membrane-bound protein kinase that phosphorylates several target proteins on tyrosine residues.

Signal peptide is a small sequence of amino acids in the structure of newly synthesized proteins that determine the eventual location of the protein in the cell.

Signaling transducers and activators of transcription (Stats) make up a group of transcription factors (seven are recognized) that are sequestered in the cytoplasm until activated by cytokine or growth factor receptors. Ligand binding causes aggregation of receptor subunits and initiation of a cascade of tyrosine phosphorylation events during which receptor-linked Jaks become activated to cause phosphorylation of the receptor. This creates a docking or binding site for Stat that is in turn phosphorylated by the receptor. Activated Stat dissociates from the receptor, dimerizes, and translocates to the nucleus for interaction with promoters of target genes. For example, Stat-5 is especially significant in the mammary gland as a mediator of many of the Prl effects on milk protein synthesis.

Somatomedin hypothesis refers to the older idea that many of the biological effects of growth hormone are mediated by GH induction of IGF-I in the liver. When first isolated, IGF-I and IGF-II were known as somatomedin A and B.

Somatostatin is a 14 amino acid peptide produced in the hypothalamus and other brain areas as well as the pancreas and gut tissue. Primarily known for its role in inhibiting secretion of growth hormone, it is also likely involved in gut-nervous system interactions.

Stroma refers to the support elements of a tissue or organ. For example, in the mammary gland this would include the extracellular proteins and various cells that surround the epithelium.

Suppressors of cytokine signaling (SOCS) are intracellular proteins that regulate activity of intracellular signaling pathways, especially Stat-associated pathways.

T

Transforming growth factor β (TGF-β) is one of a number of structurally related growth factors involved in regulation of growth, differentiation, and development in many organ systems. Three specific isoforms are expressed in the mammary gland of rodents.

V

Vagotomy indicates the severing or cutting of the vagus nerve or one of its branches.

X

Xenotransplantation indicates the transplantation of tissue or organs between unrelated species.

Index

in milk, 82, 121
periparturient levels of circulating nonesterified,
    221–222, 230
synthesis, 93–94, 95–96
Fetal mammary gland development, 22–24
Fibroblast growth factor (FGF), 157–159
FIL (feedback inhibitor of lactation), 73
Food consumption. *See* Voluntary dry matter
    intake (VDMI)
Furstenberg's rosette, 109, 110

G
G proteins, 59–62
Galactopoiesis
    endocrine regulation of, 172–198
    glucocorticoids, 177–178
    growth hormone, 181–195
    insulin, 179
    insulin-like growth factors, 195–196
    milk removal effect on, 197–198
    milking and suckling, hormone secretion with,
        173–174
    oxytocin, 178–179
    pregnancy and, 196–197
    prolactin, 174–177
    thyroid hormones, 179–181
Gestation, mammary development during, 29–35
Glucagon, 228
Glucagon-like peptide (GLP), 228–229
Glucocorticoid-binding globulin (CBG), 178
Glucocorticoids
    galactopoiesis, 177–178
    mammary development and, 144
    milk yield and, 178
    periparturient profile, 168, 169
    stress and, 237
Gluconeogenesis, 90
Glucose metabolism, milk biosynthesis and,
    90–91
Glycolysis, 90–91
Growth factors
    in mammary development, 144–162
        epidermal growth factor (EGF), 151–156
        fibroblast growth factor (FGF), 157–159
        hepatocyte growth factor (HGF), 161–162
        parathyroid hormone-related peptide, 163
        transforming growth factor (TGF), 144–145,
            159–161
    in milk, 210–215
Growth hormone, 62–64. *See also* Bovine soma-
    totropin (bST)
    concentration during lactation, 223
    galactopoiesis, 181–195

insulin-like growth factor and, 251–252
mammary development and, 129, 145–146,
    250–252
in milk, 211, 212
photoperiod affect on secretion, 241–242
thyroid hormone relationship, 181
Gut closure, 201

H
Heparin-like glycosaminoglycans, 157
Hepatocyte growth factor (HGF), 161–162
Histology, of mammary development, 9, 11, 12,
    16–17
Homeorrhesis, 45, 220
Hormones. *See also* Endocrine system; specific
    hormones
    classification of, 57–59
    concentration changes during pregnancy and
        lactation, 224
    mechanisms of action, 59–65
    in milk, 210–215
    radioimmunoassay (RIA), 130–131
    voluntary dry matter intake (VDMI), affect on,
        225–230
Hydroxyproline, 15
Hypocalcemia, parturient, 231–232
Hypothalamus, 57–59

I
Immunity. *See also* Colostrum; Immunoglobulins
    colostrum and, 199–201
    overview, 201–202
Immunoglobulins, 8
    classes of, 202
    in colostrum, 199, 200–201, 206–207
    concentration of, 202
    intestinal absorption of, 199, 201
    opsonization, 204
    structure and function, 203–204
    transpacental transfer, 199
    transport and receptors, 204–206
Insulin
    concentration during lactation, 223
    galactopoiesis, 179
    voluntary dry matter intake (VDMI) and, 228
Insulin-like growth factor-binding proteins
    (IGFBPs), 148–151, 153, 193
Insulin-like growth factor (IGF), 46, 47, 56–57
    bST administration and, 190, 192–193
    galactopoiesis, 195–196
    growth hormone and, 251–252
    in mammary development, 145–151, 152, 153
    in milk, 211–212, 214

Printed and bound by CPI Group (UK) Ltd, Croydon, CR0 4YY

16/04/2025

14658420-0003